# Sedimentology

# Sedimentology

Amber Jenkins

Larsen & Keller
www.larsen-keller.com

Sedimentology
Amber Jenkins
ISBN: 978-1-64172-642-9 (Hardback)

**☰ Larsen & Keller**

Published by Larsen and Keller Education,
5 Penn Plaza,
19th Floor,
New York, NY 10001, USA

**Cataloging-in-Publication Data**

Sedimentology / Amber Jenkins.
        p. cm.
Includes bibliographical references and index.
ISBN 978-1-64172-642-9
1. Sedimentology. 2. Petrology. 3. Sediments (Geology). I. Jenkins, Amber.
QE471 .S43 2022
552.5--dc23

For more information regarding Larsen and Keller Education and its products, please visit the publisher's website www.larsen-keller.com

# Table of Contents

Preface      VII

| Chapter 1 | Sediment and its types | 1 |
|---|---|---|
| | Marine Sediments | 4 |
| | Clastic Sediments | 7 |
| | Biogenic Sediments | 10 |
| | Chemical Sediments | 15 |

| Chapter 2 | Understanding Sedimentology | 17 |
|---|---|---|
| | Sedimentary Facies | 20 |
| | Stratigraphy | 22 |
| | Carbonate Platform | 31 |
| | Sedimentary Environments | 39 |

| Chapter 3 | Sedimentation | 44 |
|---|---|---|
| | Grain Size and Composition | 48 |
| | Fluvial Processes | 50 |
| | Aeolian Processes | 52 |
| | Glacial Deposition | 57 |

| Chapter 4 | Sedimentary Structures and Sedimentary Rocks | 62 |
|---|---|---|
| | Sedimentary Structures | 62 |
| | Soft-Sediment Deformation Structures | 63 |
| | Bedding | 64 |
| | Ripple Marks | 73 |
| | Mud Cracks | 77 |
| | Dish Structure | 80 |
| | Palaeochannel | 82 |
| | Vegetation-Induced Sedimentary Structures | 83 |
| | Hummocky Cross-Stratification | 86 |
| | Liesegang Rings | 87 |

- Sole Markings                                                    90
- Sedimentary Rocks                                                92
- Clastic Sedimentary Rocks                                        94
- Conglomerate                                                     99
- Sandstone                                                       102
- Mudrock                                                         111
- Phosphorite                                                     122
- Limestone                                                       126
- Evaporite                                                       135
- Flint                                                           141
- Iron-Rich Sedimentary Rocks                                     146
- Chert                                                           152
- Dolomite                                                        156
- Carbonate Rock                                                  157

**Chapter 5  Sedimentary Basin**                                  **160**
- Formation of Sedimentary Basin                                  163
- Classification of Sedimentary Basins                            166
- Sedimentary Isostasy                                            174
- Sedimentary Basin Analysis                                      183
- Pull-Apart Basin                                                184
- Sedimentary Basins and Petroleum Systems                        188

**Chapter 6  Uses of Sediments**                                  **189**
- Real-Life Applications of Sediments                             189
- Oil/Fuel                                                        192
- Coal                                                            193
- Uses of Sedimentary Rocks in Civil Engineering                  194

**Permissions**

**Index**

# Preface

This book is a culmination of my many years of practice in this field. I attribute the success of this book to my support group. I would like to thank my parents who have showered me with unconditional love and support and my peers and professors for their constant guidance.

The study of sediments like silt, sand and clay is referred to as sedimentology. It also studies the processes which result in their formation, transportation, deposition and diagenesis. The main types of sedimentary rocks which are studied in this field include carbonates, evaporites and clastics. Carbonates are made up of various carbonate minerals such as calcium carbonate. The evaporation of water at the surface of the Earth results in the formation of evaporites. The particles that are derived from the weathering and erosion of precursor rocks make clastic rocks. There are numerous methods used within this field to gather data and evidence on the nature and depositional conditions of sedimentary rocks. A few of them are sequence stratigraphy, describing the lithology of the rock and analyzing the geochemistry of the rock. The topics included in this book on sedimentology are of utmost significance and bound to provide incredible insights to readers. It aims to shed light on some of the unexplored aspects of this field. This book attempts to assist those with a goal of delving into this field.

The details of chapters are provided below for a progressive learning:

Chapter – Sediment and its types

Sediment refers to the naturally occurring materials which are broken down by the processes of erosion and weathering and are transported by the action of water, wind or ice. There are various types of sediments such as marine sediments, clastic sediments, biogenic sediments and chemical sediments. This is an introductory chapter which will introduce briefly all these types of sediments.

Chapter – Understanding Sedimentology

The study of sediments such silt, sand and clay is referred to as sedimentation. It also deals with the study of the processes that result in their formation such as weathering and erosion, transport and deposition. The chapter closely examines the key concepts of sedimentology to provide an extensive understanding of the subject.

Chapter – Sedimentation

The tendency of particles in suspension to settle out of the fluid in which they are entrained and come to rest against a barrier is known as sedimentation. Some of the processes which are studied in relation to the process of sedimentation are fluvial processes and aeolian processes. This chapter has been carefully written to provide an easy understanding sedimentation and related processes.

Chapter – Sedimentary Structures and Sedimentary Rocks

Rocks that are formed by the deposition or accumulation of small particles are known as sedimentary rocks. The large three dimensional physical features of sedimentary rocks are known as sedimentary structures. This chapter discusses in detail various types of sedimentary structures and rocks as well as their features.

Chapter – Sedimentary Basin

Sedimentary basin refers to a depression in the Earth's crust that is formed by plate tectonic activity in which sediments accumulate. They can be classified on the basis of the tectonic setting and geometry palaeography. The chapter closely examines the classifications of sedimentary basins to provide an extensive understanding of the subject.

Chapter – Uses of Sediments

There are various uses of sediments. They play an important role in the formation of soil which is essential for growing crops. Sedimentary rocks are used in civil engineering for building construction, cement production, pavement and road production, etc. The topics elaborated in this chapter will help in gaining a better perspective about these diverse uses of sediments.

**Amber Jenkins**

# Sediment and its types

Sediment refers to the naturally occurring materials which are broken down by the processes of erosion and weathering and are transported by the action of water, wind or ice. There are various types of sediments such as marine sediments, clastic sediments, biogenic sediments and chemical sediments. This is an introductory chapter which will introduce briefly all these types of sediments.

Sediment is any particulate matter that is transported by the flow of fluids (such as water and air) and eventually deposited in a layer of solid particles. The process of deposition by settling of a suspended material is called sedimentation.

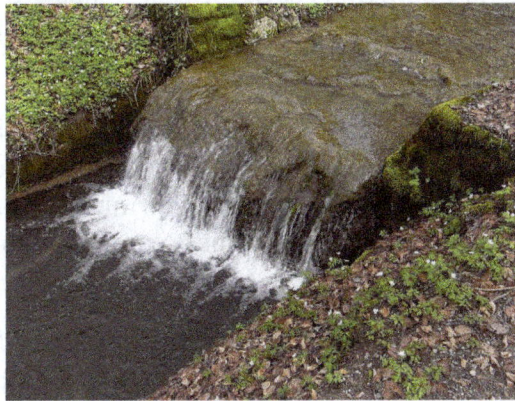

Sediment builds up on human-made breakwaters because they reduce the speed of water flow, so the stream cannot carry as much sediment load.

Sediments may be transported by the action of streams, rivers, glaciers, and wind. Desert sand dunes and loess (fine, silty deposits) are examples of eolian (wind) transport and deposition. Glacial moraine (rock debris) deposits and till (unsorted sediment) are ice-transported sediments. In addition, simple gravitational collapse, as occurs after the dissolution of layers of bedrock, creates sediments such as talus (slope formed by accumulated rock debris) and mountainslide deposits.

Seas, oceans, and lakes also accumulate sediment over time. The material can be terrestrial (deposited on the land) or marine (deposited in the ocean). Terrigenous deposits originate on land and are carried by rivers and streams, but they may be deposited in either terrestrial, marine, or lacustrine (lake) environments. In the mid-ocean, living organisms are primarily responsible for sediment accumulation, as their shells sink to the ocean floor after the creatures die.

The process of sedimentation helps renew nutrients in the soil, thereby supporting living organisms. Without such processes, the soil could become depleted of nutrients relatively quickly, and living organisms may not be able to survive in those same habitats. Moreover, deposited sediments are the source of sedimentary rocks, which can contain fossils that were covered by accumulating sediment. Lake-bed sediments that have not solidified into rock can be used to determine past climatic conditions. Thus, by analyzing sediments and sedimentary rocks, we can get glimpses of some aspects of the Earth's history.

## Key Depositional Environments

## Fluvial Bedforms

Rivers and streams are known as fluvial environments. Any particle that is larger in diameter than approximately 0.7 millimeters will form visible topographic features on the riverbed or streambed. These features, known as bedforms, include ripples, dunes, plane beds, and antidunes. The bedforms are often preserved in sedimentary rocks and can be used to estimate the direction and magnitude of the depositing flow.

The major fluvial environments for deposition of sediments include the following:

1. Deltas: River deltas, which are arguably intermediate between fluvial and marine environments, are landforms created by the buildup of sediment at the "mouths" of rivers and streams, that is, at places where they reach the sea. Deltas are roughly triangular in shape, but the shape depends on how the water flows, how the current changes, and the amount of sediment being carried.

2. Point bars: They are the result of an accumulation of gravel, sand, silt, and clay on the inside bank of a bend of a river. They demonstrate a characteristic semi-ellipse shape because of the way they are formed, with larger sediment forming the base, and finer particles making up the upper part of the point bar. Point bars contribute to size and shape changes of a meander (bend) over time.

3. Alluvial fans: These are fan-shaped deposits formed where a fast-flowing stream flattens, slows, and spreads, typically at the end of a canyon onto a flatter plain.

4. Braided rivers: They consist of a network of small channels separated by small and often temporary islands called *braid bars*. Braided streams are common wherever a drastic reduction in stream gradient causes rapid deposition of the stream's sediment load.

5. Oxbow lakes: These are curved lakes formed when a wide meander (or bend) of a nearby stream or river is cut off. A combination of deposition and rapid flow work to seal the meander, cutting it off from the original body of water it was formerly connected to.

6. Levees: These are natural or artificial embankments or dikes that border the perimeter of a river. They have a wide earthen base and taper at the top. Natural levees occur as a result of tidal waves or sharp meandering of a river. Artificial levees are built to prevent flooding of the adjoining land, but they also confine the river flow, increasing the velocity of the flow.

## Marine Bedforms

Marine environments (seas and oceans) also see the formation of bedforms. The features of these bedforms are influenced by tides and currents. The following are major areas for deposition of sediments in the marine environment.

1. Littoral (coastal) sands: They include beach sands, coastal bars and spits. They are largely clastic, with little faunal content.

2. The continental shelf: It consists of silty clays, with increasing content of marine fauna.

3. The shelf margin: It has a low supply of terrigenous material, mostly faunal skeletons made of calcite.

4. The shelf slope: This consists of much more fine-grained silts and clays.

5. Beds of estuaries: The resultant deposits are called "bay mud."

One other depositional environment, called the turbidite system, is a mixture of fluvial and marine environments. It is a major source of sediment for the deep sedimentary and abyssal basins, as well as for deep oceanic trenches.

## Surface Runoff

Surface runoff water can pick up soil particles and transport them in overland flow for deposition at a lower land elevation or deliver that sediment to receiving waters. In this case, the sediment is usually deemed to result from erosion. If the initial impact of rain droplets dislodges soil, the phenomenon is called "splash erosion." If the effects are diffuse for a larger area and the velocity of moving runoff is responsible for sediment pickup, the process is called "sheet erosion." If there are massive gouges in the earth from high-velocity flow for uncovered soil, then "gully erosion" may result.

## Rate of Sediment Settling

When a fluid (such as water) carries particles in suspension, the process by which the particulates settle to the bottom and form a sediment is called settling. The term settling velocity (or fall velocity or terminal velocity (ws)) of a particle of sediment is the rate at which the particle settles in still fluid. It depends on the size, shape, and density of the grains, as well as the viscosity and density of the fluid.

For a dilute suspension of small, spherical particles in a fluid (air or water), the settling velocity can be calculated by Stoke's Law:

$$w = \frac{2(\rho_p - \rho_f)gr^2}{9\mu}$$

where $w$ is the settling velocity; $\rho$ is density (the subscripts $p$ and $f$ indicate particle and fluid respectively); $g$ is the acceleration due to gravity; $r$ is the radius of the particle; and $\mu$ is the dynamic viscosity of the fluid.

If the flow velocity is greater than the settling velocity, sediment will be transported downstream as suspended load.

As there will always be a range of different particle sizes in the flow, some will have sufficiently large diameters that they settle on the riverbed or streambed but still move downstream. This is known as *bed load*, and the particles are transported via such mechanisms as rolling, sliding, and "saltation". Saltation marks are often preserved in solid rocks and can be used to estimate the flow rate of the rivers that originally deposited the sediments.

## Erosion

One of the main causes of riverine sediment load siltation stems from "slash and burn" treatment of tropicalforests. When the ground surface is stripped of vegetation and seared of all living organisms, the upper soils are vulnerable to both wind and water erosion. In a number of parts of the world, entire sectors of a country have been rendered erosive.

For example, on the Madagascar high central plateau, comprising approximately ten percent of that country's land area, virtually the entire landscape is devoid of vegetation, with gully erosive furrows typically in excess of 50 meters deep and one kilometer wide.

Shifting cultivation is a farming system that sometimes incorporates the slash and burn method in some areas of the world. The resulting sediment load in rivers is ongoing, with most rivers a dark red brown color. The accumulation of these fine particulates in the water also lead to massive fish kills, as they cover fish eggs along the bottom floor.

# MARINE SEDIMENTS

Marine sediment is any deposit of insoluble material, primarily rock and soil particles, transported from land areas to the ocean by wind, ice, and rivers, as well as the remains of marine organisms, products of submarine volcanism, chemical precipitates from seawater, and materials from outer space (e.g., meteorites) that accumulate on the seafloor.

Although systematic study of deep-ocean sediments began with the HMS *Challenger* expeditions between 1872 and 1876, intensive research was not undertaken until nearly 100 years later. Since 1968 American scientists, in collaboration with those from the United Kingdom, the Soviet Union, and various other countries, have recovered numerous sedimentary core samples from the Atlantic and Pacific oceans through the use of a specially instrumented deep-sea drilling vessel called the *Glomar Challenger*.

Marine sediments deposited near continents cover approximately 25 percent of the seafloor, but they probably account for roughly 90 percent by volume of all sediment deposits. Submarine canyons constitute the main route for sediment movement from continental shelves and slopes onto the deep seafloor. In most cases, an earthquake triggers a massive slumping and stirring of sedimentary material at the canyon head. Mixed with seawater, a dense liquid mass forms, giving rise to a density current that flows down the canyon at speeds of several tens of kilometres per hour. After reaching the base of the continental slope, the sediment-laden mass moves out onto the continental rise at the base of the slope. Deposits from turbidity currents (i.e., short-lived density currents caused by suspended sediment concentrations) can build outward for hundreds and sometimes thousands of kilometres across the ocean bottom. Large sediment-built plains commonly occur in the Atlantic Ocean, where turbidity currents flow from the base of a continent to the Mid-Atlantic Ridge.

Deposits produced by turbidity currents are called turbidites. Most of them consist of sands and silts, but a few are composed of gravels. Turbidites tend to have distinct boundaries between adjacentunits. Each of these units is formed by a separate flow and often exhibits a systematic change in grain size from coarsest at the bottom to finest at the top. Turbidites characteristically contain the remains of shallow-water organisms mixed with deep-water varieties. The shallow-water organisms came from areas where the density current originated, whereas the deep-water forms existed in the area traversed by the current or where it finally deposited its load.

The sediments deposited on continental shelves and rises, frequently referred to as hemipelagic sediments, ordinarily accumulate too rapidly to react chemically with seawater. In most cases, individual grains thus retain characteristics imparted to them in the area where they formed. As a rule, sediments deposited near coral reefs in shallow tropical waters contain abundant carbonate material. Calcareous, reef-derived muds, for example, occur around atolls at the northwestern end of the Hawaiian Island chain. Near volcanoes, sediments contain ash—e.g., silicate glass and fine volcanic-rock fragments.

Roughly 75 percent of the deep seafloor is covered by slowly accumulating deposits known as pelagic sediments. Because of its great distance from the continents, the abyssal plain does not receive turbidity currents and their associated coarse-grained sediments. Moreover, since relatively little land-derived sediment consisting of silicate mineral and rock fragments reach the ocean bottom, deposits there show a predominance of biogenic constituents (i.e., the skeletal remains of marine organisms). In areas where surface waters are fertile, opal from diatoms (algae) and radiolarians (protozoans) and calcium carbonate from such organisms as foraminiferans, coccolithophorids, and pteropods are supplied to the sediment. If the biological constituents exceed 30 percent by volume, then the deep-ocean sediments are usually classified on the basis of their biogenic components. For example, a mud containing 30 percent by volume of foraminiferal tests (external hard parts) is called a foraminiferal mud or ooze. When one genus dominates, it is frequently referred to by the generic name, such as *Globigerina* ooze. Diatomaceous and radiolarian muds are named on the same basis. Where biogenic constituents compose less than 30 percent of the total, the deposit is called a deep-sea clay, brown mud, or red clay.

The deep-ocean bottom is continually renewed through seafloor spreading . Oceanic crust is created at the mid-oceanic ridges as a consequence of extrusive igneous activity and moves away, carrying along overlying sediments. Over time, the crust and the associated sedimentary material are destroyed at the oceanic trenches. The sedimentary core samples recovered by the "Glomar Challenger" strongly support the seafloor-spreading hypothesis. No deep-sea sediments older than 150,000,000 years were discovered, indicating that the seafloor is relatively young. Furthermore, the sediments become progressively older and thicker with increasing distance from the ridge crests.

## Types of Marine Sediment

Sea floors are mostly covered in sediments which are formed by various sources with different compositions. Marine sediments are either organic or inorganic particles consisting of rock and soil particles, remains of marine organisms, volcanoes, plants, animals and outer space. Marine sediments are classified according to their origin. There are four common types of marine sediments – Lithogenous, Biogenous, Hydrogenous, and Cosmogenous.

Distribution of marine sediments in the world map.

## Lithogenous

Lithogenous sediments are derived from land thus their sources are the rocks of the earth's crust. Lithogenous sediments are produced through weathering process and are made up of small weathered rocks and volcanic activity. They are transported from land by water, wind, ice, gravity etc. They are also commonly called terrigenous sediments and are the most abundant type of sediments on earth. Lithogenous sediments are often formed when metal and silicate ions bond together. Most lithogenous materials get concentrated continental margins while others get deposited into the deep ocean.

There are two categories of lithogenous sediments: Terrigenous and redclay. Terrigenous sediments are produced by the weathering of rocks above the water. The eroded particles are transported by wind and other natural means and get deposited at the bottom of oceans. Red clay on the other hand is abundant in the ocean. They are reddish brown in colour and a combination of terrigenous material and volcanic ash. They are transported by currents and winds and get deposited along the ocean floor.

## Biogenous

Biogenous sediments are formed from the remains of living organisms such as bones, teeth, shells of animals, terrestrial and aquatic plants and organic residues. They are one of the most important constituents of ocean sediments and cover around 75% of the seafloor. Calcium carbonate ($CaCO_3$) and silica ($SiO_2$) are the two most common chemical compounds found in this sediment. Calcareous sediments contain foraminifera, coccolithophores and pteropods protective coverings, while siliceous sediment consists diatom and radiolaria protective coverings. Common organisms that are primarily found here are skeletons and shells of marine organisms and microscopic plankton. Most of these sediments are biologically produced inorganic matter, and only a small section is organic matter.

## Hydrogenous

Hydrogenous sediments occur as a result of chemical reaction within the sea water. They are found on deep-ocean floors containing Iron oxides and manganese. These sediments accumulate very slow and make up less than 1% of ocean sediments. Manganese, phosphorite and carbonates are common hydrogenous sediments. In shallow seas, rock salt, sulfates and calcium salts may be found on the ocean floor.

## Cosmogenous

Cosmogenous sediments are meteoric debris falling from outer space. They are the remains from the impact of large bodies of space like comets or other bodies which land and settle on the ocean floor. These sediments are composed of silicates and mixtures of different metals and are very rare.

# CLASTIC SEDIMENTS

Clastic sediment is sediment consisting of fragments of rock, transported from elsewhere and re-deposited to form another rock. Clasts are individual grains that make up the sediments. The sediment particles are then further exposed to rain, wind, and gravity, which batters and break them apart through further weathering and erosion processes.

The products of weathering will finally include particles ranging from clay to silt, to pebbles and boulders, that are then suspended and transported downstream by wind, streams, rivers, and ocean tides and currents to the earth's ocean and sea basins below, where they are buried, lithified, subjected to heat and pressure at various depths to solidify into the many different sedimentary rock types that exist.

As the earth consists 70% of water, a great majority of sediments will form into the estuaries, deltas, seas, lakes, and oceans to form sedimentary sequences that will often result in kilometers of sedimentary rock sequences below the subsurface, i.e., seabed, where, when deep enough, further pressure, heat, and temperature changes further cook and change the sedimentary rock.

Above the metamorphic bedrocks within the earth basins, sediment thicknesses overlying the majority of the world's oceans, seas, and margins have been mapped, interpreted, and can be readily obtained to conclude deepwater sedimentary basin sequences and rock thickness where hydrocarbons exist are not all the same.

Clastic sediments are predominantly clay minerals and quartz particles, with minor amounts of Feldspars, micas, and heavy minerals. Porosity results from the space between the grain particles that is not filled with cement or clay. Porosity is usually in the range from 10% to 30% depending on the grain sizes, compaction, and the amount of cement present between the pores. Permeability, which is the property that permits fluid to flow through the pores, is controlled by the amount of cement, the degree of compaction, and the magnitude and variation of grain sizes.

Clastic sediments predominate under cold climatic conditions, such as those found in the Arctic or in high Alpine regions . Such sediments are typical for proglacial and periglacial lakes. Intensive physical weathering and the lack of a densely vegetated catchment area provide high amounts of minerogenic detritus, which is easily eroded and transported into the lake. The sediment transfer is related to the annual freeze–thaw cycle and the amount of runoff. In regions with a continental climate, runoff is governed by melting of snow and ice through solar insolation during summer. Under oceanic climatic conditions, runoff is controlled either by the melting of snow and ice through advective heat transport or by precipitation. Such lakes are poor in nutrients (oligotrophic), which inhibits high organic productivity and the formation of organic varves.

Clastic varves of Lake C$_2$, north coast of Ellesmere Island, Canadian High Arctic. The microscopic image of nonglacial clastic varves shows the undulated sediment–water interface in normal (left) and polarized light (right). Pale and coarse-grained laminae are related to snowmelt runoff events during summer, which contrast to the dark fine-grained laminae on top that settled out of suspension during winter under ice cover.

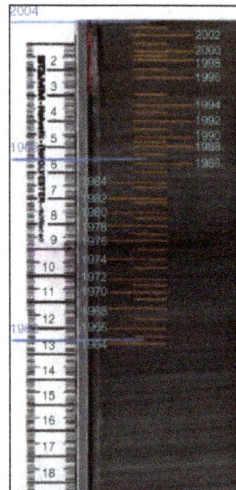

Nonglacial clastic varves of Alpine Brienzersee, Switzerland. Varve counts are marked and labeled; the scale is in cm.

Clastic varves are formed when a discontinuous (highly seasonal) stream loaded with suspended sediment enters a lake of stratified water. According to its density and in relation to the density of the lake water, the inflowing stream water is positioned in the lake as over-, inter-, or underflow. Over- and interflows cause a wide distribution of suspended matter across the lake. After entering the lake, the flow velocity of the stream water is transformed into turbulence causing a reduction of transport capacity for suspended sediment particles. Accordingly, coarse particles (sand, silt) are deposited immediately, whereas fine particles (fine silt, clay) remain in suspension until runoff has stopped or the lake is ice-covered. Thus, clastic varves are composed of a pale coarse-grained basal lamina and a darker fine-grained (clay) top lamina or vice versa. Additional coarse-grained laminae are frequently observed and are often related to successive annual runoff events.

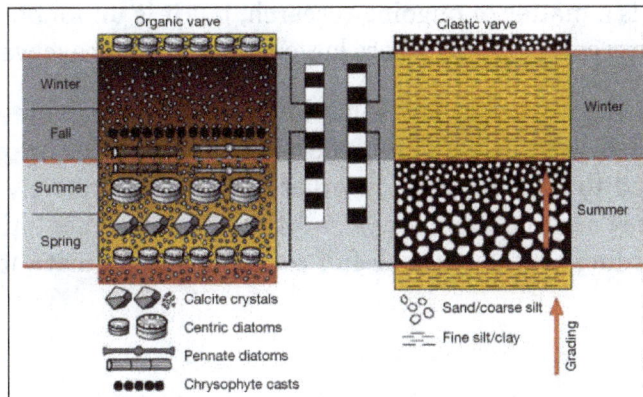

Model of varve formation for carbonaceous organic varves (left) and clastic varves (right).

Table: Different varve types with idealized composition, color, and timing of corresponding seasonal laminae.

| Clastic varves | Organic varves | Evaporitic varves | |
|---|---|---|---|
| Spring | | First diatom bloom (pale) | |
| Summer | Coarse-grained minerogenic particles via runoff (pale or dark) | Calcite precipitation (white, optional) | Precipitation of calcite, aragonite, gypsum or halite (white) |
| Late summer | | Second diatom bloom, chrysophytes cysts (pale, both optional) | |
| Fall/winter | Fine-grained minerogenic particles out of suspension (dark or pale) | Organic and minerogenic detritus via runoff (dark) | Organic and minerogenic detritus via runoff (dark) |

## Sediment Production

Clastic sediments are produced by the physical disaggregation of preexisting rocks during weathering and mechanical erosion. Chemical weathering weakens rocks by altering mineral compositions and by removing the minerals cementing them together. Physical weathering involves fragmentation due to tensional or compressive stresses, including expansion of clay minerals within rocks, frost shatter, and root wedging. The latter two are frequently sources of angular rock fragments, including breakdown, in and near cave entrances. Breakdown blocks are clastic sediment and they easily exceed the total volume of all other clastic deposits in a cave, occurring where passage widths exceed the tensile strength of ceiling or overhanging rock beds.

Cave streams mechanically erode conduit walls through a combination of plucking and corrasion. Plucking adds sediment to karst streams when intense hydraulic forces pull or wedge blocks of rock from conduit surfaces. Corrasion occurs when sediment grains carried by floodwaters strike bedrock surfaces with sufficient force to break off fragments. This process is generally referred to as abrasion when the result is a smooth or polished cave surface and percussion when it results in obvious flaking and rock breakage. Percussion generally implies impact of large grains, such as cobbles, whereas abrasion is most commonly associated with "sand blasting" by suspended sediment. Solid carbonate grains created by abrasion can collectively weigh more than the dissolved solids in those same cave-fed floodwaters. Whether corrasion plays a major role in cave development elsewhere is a matter of ongoing research, but it is undoubtedly a nontrivial source of sediment to high-energy cave streams where large grains impact cave walls.

Carbonates may contain chert or other low-solubility minerals, which enter cave passages as the surrounding rock is eroded. Where abundant, chert can be the major or only significant autochthonous bedload supplied to cave streams. In other cases, comparatively high concentration of autochthonous grains may accumulate in clays and silts when sedimentation rates are otherwise very slow, perhaps recording protracted pipe-full conditions in a quietwater setting (many hundreds of years).

# BIOGENIC SEDIMENTS

Biogenic Sediments are defined as containing at least 30% skeletal remains of marine organisms, cover approximately 62% of the deep ocean floor. Clay minerals make up most of the non-biogenic constituents of these sediments. While a vast array of plants and animals contribute to the organic matter that accumulates in marine sediments, a relatively limited group of organisms contribute significantly to the production of biogenic deep-sea sediments, which are either calcareous or siliceous oozes.

Distributions and accumulation rates of biogenic oozes in oceanic sediments depend on three major factors:

- Rates of production of biogenic particles in the surface waters.

- Dissolution rates of those particles in the water column and after they reach the bottom.

- Rates of dilution by terrigenous sediments.

The abundances and distributions of the organisms that produce biogenic sediments depend upon such environmental factors as nutrient supplies and temperature in the oceanic waters in which the organisms live. Dissolution rates are dependent upon the chemistry of the deep ocean waters through which the skeletal remains settle and of the bottom and interstitial waters in contact with the remains as they accumulate and are buried. The chemistry of deep-sea waters, is, in turn, influenced by the rate of supply of both skeletal and organic remains of organisms from surface waters. It is also heavily dependent upon the rates of deep ocean circulation and the length of time that the bottom water has been accumulating $CO_2$ and other byproducts of biotic activities.

## Carbonate Oozes

## Production

Most carbonate or calcareous oozes are produced by the two different groups of organisms. The major constituents of nanofossil or coccolith ooze are tiny (less than 10 microns) calcareous plates produced by phytoplankton of the marine algal group, the Coccolithophoridae or by an extinct group called discoasters. Foraminiferal ooze is dominated by the tests (shells) of planktic protists belonging to the Foraminiferida. Most foraminiferal tests are sand-sized (>61 mm in diameter), so many foraminiferal oozes are bimodal in particle-size distribution, because they are made up of sand-sized foraminiferal tests and mud-sized coccolith plates.

Discoasters, coccoliths and foraminiferal tests are all made of the mineral calcite. Pteropod ooze is produced by the accumulation of shells of pteropods and heteropods, which are small planktic mollusks. As these shells are composed of the mineral aragonite, pteropod oozes are more easily dissolved, so are restricted to relatively shallow depths (less than 3,000 m) in tropical areas.

Carbonate oozes are the most widespread shell deposits on earth. Nearly half the pelagic sediment in the world's oceans is carbonate ooze. Furthermore, foraminifera and coccolithophorids have been major producers of pelagic sediment for the past 200 million years. As a result, these are arguably among the most important and scientifically useful organisms on Earth. Because their larger size makes them easier to identify and work with, this is particularly true for the foraminifera. Their fossils provide the single most important record of Earth history over the past 200 million years. That history is recorded not only by the evolution of species and higher taxa through that time, but is also preserved in the chemistry of the fossils themselves. The field of Paleoceanography owns much of its existence to biostratigraphy, isotope stratigraphy and paleoenvironmental analyses that utilize fossil foraminifera.

The distributions and abundances of living planktic foraminifera and coccolithophorids in the upper few hundred meters of the ocean depends in large part on nutrient supply and temperature. Coccolithophorids, because they are marine algae, require sunlight and inorganic nutrients (fixed N, P, and trace nutrients) for growth. However, most coccolithophorid species grow well with very limited supplies of nutrients and do not compete effectively with diatoms and dinoflagellates when nutrients are plentiful. Furthermore, both high nutrient supplies and cold temperatures inhibit calcium carbonate production to some degree. For these reasons, diversities (number of different kinds) of coccolithophorids are high and production rates of coccoliths are moderate even in the most nutrient-poor regions of the subtropical oceans, the subtropical gyres. Production of coccoliths is higher in equatorial upwelling zones and often along continental margins and in temperate latitudes where nutrient supplies are higher, though diversities decline. In very high nutrient areas, such as upwelling zones in the eastern tropical oceans (*i.e.*, meridional upwelling), polar divergences and near river mouths, production of coccoliths is minimal.

Even though planktic foraminifera are protozoans rather than algae, their distributions, diversities, and carbonate productivity are quite similar to those of coccolithophorids. Many planktic foraminifera, especially the spinose species that live in the upper 100 m of temperate to tropical oceans host dinoflagellate symbionts which aid the foraminifera by providing energy and enhancing calcification. Having algal symbionts is highly advantageous in oceanic waters where inorganic nutrients and food are scarce, so a diverse assemblage of planktic foraminifera thrives along with the coccolithophorids in the nutrient-poor subtropical gyres. Greater abundances of fewer species thrive in equatorial upwelling zones and along continental margins, so rates of carbonate shell production are higher. And similar to coccolithophorids, few planktic foraminifera live in very high nutrient areas, such as upwelling zones in the eastern tropical oceans, polar divergences and near river mouths, so production of carbonate sediments is minimal in these areas. Finally, planktic foraminifera require deep oceanic waters to complete their life cycles, which they cannot do in neritic waters over continental shelves.

Cool temperatures work together with higher nutrient supplies to reduce diversities of coccolithophorids and planktic foraminifera, and ultimately to shift the ecological community to organisms that do not produce carbonate sediments. A 10 °C drop in temperature is physiologically similar to doubling nutrient supply, which is why the pelagic community in an equatorial upwelling zone resembles that of a temperate oceanic region, while the pelagic community of an intensive meridional upwelling zone resembles subpolar to polar communities.

If surface production was the only factor controlling accumulation rates of carbonate oozes, deep-sea sediment patterns would be quite simple. Carbonate oozes would cover the seafloor everywhere except

- Beneath intensive meridional upwelling zones,

- Beneath polar seas,

- Where they are overwhelmed by terrigenous sedimentation.

Rates of accumulation would be on the order of 3-5 cm/1000 years in the open ocean and 10-20 cm/year beneath equatorial upwelling zones and along most continental margins.

## Dissolution

Over much of the ocean floor, carbonate accumulation rates are controlled more by dissolution in bottom waters than by production in surface waters. Dissolution of calcium carbonate in seawater is influenced by three major factors: temperature, pressure and partial pressure of carbon dioxide ($CO_2$). The easiest way to understand calcium carbonate ($CaCO_3$) dissolution is to recognize that it is controlled, in large part, by the solubility of $CO_2$:

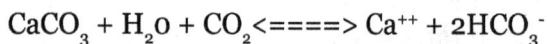

$$CaCO_3 + H_2O + CO_2 <====> Ca^{++} + 2HCO_3^-$$

The more $CO_2$ that can be held in solution, the more $CaCO_3$ that will dissolve. Since more $CO_2$ can be held in solution at higher pressures and cooler temperatures, $CaCO_3$ is more soluble in the deep ocean than in surface waters. Finally, as $CO_2$ is added to the water, more $CaCO_3$ can dissolve. The result is that, as more $CO_2$ is added to deep ocean water by the respiration of organisms, the more corrosive the bottom water becomes to calcareous shells.

The rain of organic matter from surface waters through time increases the partial pressure of $CO_2$ in bottom water, so the longer the bottom water has been out of contact with the surface, the higher its partial pressure of $CO_2$. Beneath high-nutrient surface waters, primary production exceeds what is utilized in the surface mixed layer. Excess organic matter falling through the water column accumulates on the bottom, where organisms feed upon it and oxidize it to $CO_2$.

The depth at which surface production of $CaCO_3$ equals dissolution is called the calcium carbonate compensation depth(CCD). Above this depth, carbonate oozes can accumulate, below the CCD only terrigenous sediments, oceanic clays, or siliceous oozes can accumulate. The calcium carbonate compensation depth beneath the temperate and tropical Atlantic is approximately 5,000 m deep, while in the Pacific, it is shallower, about 4,200-4,500 m, except beneath the equatorial upwelling zone, where the CCD is about 5,000 m. The CCD in the Indian Ocean is intermediate between the Atlantic and the Pacific. The CCD is relatively shallow in high latitudes.

Surface waters of the ocean tend to be saturated with respect to $CaCO_3$; low latitude surface waters are usually supersaturated. At shallow to intermediate seafloor depths (less than 3000 m), foraminiferal tests and coccolith plates tend to be well preserved in bottom sediments. However, at depths approaching the CCD, preservation declines as smaller and more fragile foraminiferal tests show signs of dissolution. The boundary zone between well preserved and poorly preserved foraminiferal assemblages is known as the lysocline.

The preservation potential of the various kinds of carbonate shells and skeletons differs. Pteropod shells are aragonite, a less stable form of $CaCO_3$. Pteropod shells dissolve at depths greater than 3,000 m in the Atlantic Ocean and below a few hundred meters in the Pacific. Calcitic planktic foraminiferal tests, especially small tests of juvenile spinose foraminifera, dissolve more readily than coccoliths, which are also made of calcite. Pelagic sediments from relatively shallow depths in low latitudes are often dominated by pteropods shells, at intermediate depths by foraminiferal tests, below the lysocline and above the CCD by coccoliths, and below the CCD by red clays.

Regional changes in the depths of the lysocline and CCD result, in part, from changes in $CO_2$ content of bottom waters as they "age". In modern oceans, deep ocean circulation is driven by formation of bottom waters during the freezing of sea ice. Seawater, due to its salt content, can cool below -1 °C before ice begins to form. When sea ice forms, the salt is excluded and is left behind in the seawater. Water in the vicinity of the freezing sea ice becomes more saline and therefore more dense. As a result, large-scale sea ice formation creates very dense water masses that sink to the bottom of the ocean to form deep bottom water. During the Antarctic winter, the freezing of sea ice in the Weddell Sea produces Antarctic Bottom Water (AABW), which sinks to the sea bottom and spreads northward into the South Atlantic. During the Arctic winter, sea ice formation in the Norwegian and Greenland Seas produce North Atlantic Deep Water(NADW), which sinks to the bottom of the North Atlantic and flows southward. AABW is slightly more dense than NADW, so when they meet, AABW flows beneath NADW. As the NADW and AABW spread eastward into the Indian and Pacific Oceans, they mix to become Deep Pacific Common Water (DPCW). The "youngest" bottom waters are in the Atlantic, the "oldest" are in the North Pacific.

When seawater is at the surface, it equilibrates with the atmosphere with respect to $O_2$ and $CO_2$. From the time a water mass sinks from the surface until it comes back to the surface, respiration by organisms in the water column and on the bottom use up $O_2$ and add $CO_2$. As a result, the longer bottom water is away from the surface, the more corrosive it is to $CaCO_3$.

## Carbonate Sedimentation Worldwide

The depth of the CCD and the pattern of carbonate sedimentation in any part of the world's ocean reflects the influences of surface production of organic matter, surface production of carbonates, and the corrosiveness of the bottom water to $CaCO_3$.

Because coccolithophorids and planktic foraminifera thrive in temperate to subtropical oceans where surface nutrient supplies are very limited, these organisms produce a continual rain of $CaCO_3$ to the sea floor. In equatorial upwelling zones, organic productivity is elevated enough to stimulate higher rates of production of calcareous and siliceous skeletal remains, but not enough to export excess organic matter to the deep ocean where its respiration would increase corrosiveness of bottom waters to $CaCO_3$.

In more intensive upwelling zones, especially in the eastern tropical Pacific and the Antarctic divergence, and off major river deltas, high nutrient supplies stimulate high rates of organic productivity by diatoms and dinoflagellates, often to the exclusion of coccolithophorids and planktic foraminifera, which reduces $CaCO_3$ production. At the same time, the rain of organic matter to the ocean floor supports abundant deep-sea life whose respiration adds significantly to the $CO_2$ in bottom waters. The result is substantial shoaling of the lysocline and CCD in these regions. The greater corrosiveness of AABW compared to NADW at approximately the same "age" is caused by upwelling-induced high organic productivity at the Antarctic divergence, which exports excess of organic matter into AABW.

Pelagic sediments in the Atlantic and Indian Oceans are predominantly calcareous oozes. In the Pacific Ocean, where the CCD is deeper, red clays dominate, especially in the North Pacific. Carbonate oozes delineate shallower regions in the south Pacific, including the East Pacific Rise and the complex topography to the southwest.

## Siliceous Oozes

Biogenic siliceous oozes have two major and two minor contributors:

- Golden-brown algae known as diatoms (Bacillariophyceae) construct a type of shell called a frustule out of opalline silica.

- The radiolaria , a large group of marine protists distantly related to the foraminifera, also construct opalline silica skeletons.

Silicoflagellates are a minor group of marine algae that also construct opalline silica skeletons. Sponge spicules are also an important biogenic source of opalline silica in neritic waters, but are of minor importance in the deep sea.

Silica is undersaturated throughout most of the world's oceans. As a result, extraction of silica from seawater for production of silica shells or skeletons requires substantial energy. Furthermore, for siliceous sediments to be preserved, they must be deposited in waters close to saturation with respect to silica and they must be buried quickly. Young seawaters that are highly undersaturated with respect to $H_4SiO_4$ are far more corrosive to $SiO_2$ than are old seawaters that have been dissolving and accumulating $H_4SiO_4$ over hundreds to thousands of years.

Seawaters around volcanic islands and island arcs tend to have higher concentrations of $H_4SiO_4$ in solution and therefore are more conducive to silica production in surface waters and silica preservation in sediments. Siliceous sediments are most common beneath upwelling zones and near high latitude island arcs, particularly in the Pacific and Antarctic. More than 75% of all oceanic silica accumulates on the sea floor between the Antarctic convergence and the Antarctic glacial marine sedimentation zone. Accumulation rates of siliceous oozes can reach 4-5 cm/1,000 years in these areas.

Conditions favoring deposition of silica or calcium carbonate are different. Silica solubility increases with decreasing pressure and increasing temperature. Silica is undersaturated in the oceans, but it is less undersaturated in deep water. Carbonate solubility increases with depth, and bottom waters become more undersaturated in calcium carbonate. The patterns of carbonate and silica deposits reflect different processes of formation and preservation, resulting in carbonate oozes that are poor in biogenic silica and vice versa.

The diatoms are extremely important primary producers that benefit physiologically from rich supplies of dissolved inorganic nutrients. Under such conditions, their growth rates far exceed other phytoplankton and they can rapidly produce both organic matter and siliceous sediments. They thrive in areas of intensive upwelling and near terrestrial sources of dissolved nutrients, including silica. Silicoflagellates show similar distributions. On the other hand, because both groups require substantial nutrient resources for growth, they are never abundant where nutrients are scarce, and so are insignificant primary and sediment producers in subtropical gyres. Diatom oozes, which contain more than 30% diatom frustules, are found beneath the Antarctic divergence, off the Aleutian island arc in the far North Pacific, and beneath areas of intensive meridional upwelling such as the eastern tropical Pacific. These oozes contain a significant percentage of radiolarian and silicoflagellate skeletons as well. Diatom-rich muds are common on continental shelves and margins where runoff from land contributes terrigenous muds as well as nutrients that stimulate diatom production.

Radiolaria, being protists, are slightly less dependent on the most nutrient-rich areas of the oceans. They are important contributors to siliceous oozes around the Antarctic, but radiolarian oozes (> 30% radiolarian skeletons) are primarily in the tropical Pacific beneath the equatorial upwelling zone and below the CCD. Above the CCD in this region, the sediments are calcareous with a significant siliceous component.

After burial, most siliceous oozes remain unconsolidated, but a fraction dissolve and reprecipitate as chert beds or nodules. Chert is cryptocrystalline and microcrystalline quartz, which is very hard and impermeable. Chert beds are very difficult to drill, which has frustrated ocean drillers since the early days of the Deep Sea Drilling Project (DSDP). The abundance and widespread distribution of chert beds of Eocene age, discovered by the DSDP, indicate important changes in deep-sea chemistry over the past 50 million years.

# CHEMICAL SEDIMENTS

Oolitic Limestone typically forms in tidal regions where particles are in constant motion.

Chemical sedimentary rock is formed when minerals, dissolved in water, begin to precipitate out of solution and deposit at the base of the water body. This can occur in hot springs, such as those in Yellowstone, where changes in water chemistry initiate precipitation of calcium carbonate in the form of travertine, or in areas where sea water evaporates, depositing rock salt or gypsum.

Due to the manner in which they are formed, these types of rocks exhibit a crystalline texture. This texture can also occur in rocks that have undergone some form of recrystallization during the lithification process. This most often occurs in rocks produced by the accumulation of siliceous or calcareous tests (or shells) of microorganisms. During the burial process, water may react within the small pores and recrystallize into fine-grained texture. Because most chemical sedimentary rocks are formed in marine environments, it is not unlikely to find fossils within chemically precipitated rocks.

Coquina. Form of limestone produced through the accumulation of shells.

In shallow marine environments a specific type of precipitation, called an ooid, can occur. In these environments material is often under constant motion due to tidal or wave action. As these materials roll back and forth on the sea floor, precipitates will crystallize on only the exposed surface. Eventually, thin spherical layers of precipitate will develop on the original particle (or nucleus). Sedimentary rocks composed of ooids are described by the precipitated mineral, for example, ooilitic limestone or ooilitic hematite.

Biochemical sedimentary rocks, also known as bioclastic sedimentary rocks, form from the gradual accumulation of biologic material such as shells or dead plant material.

## References

- Sediment, entry: newworldencyclopedia.org, Retrieved 14 August, 2019

- Marine-sediment, science: britannica.com, Retrieved 15 January, 2019

- What-are-the-four-types-of-marine-sediments, geography: desanacademy.com. Retrieved 16 February, 2019

- Clastic-sediment, earth-and-planetary-sciences, topics: sciencedirect.com, Retrieved 17 March, 2019

- 8-biogenic, morelockonline, morelocksite: geology.uprm.edu, Retrieved 18 April, 2019

- Sedimentary-Rocks-chemical: thisoldearth.net, Retrieved 19 May, 2019

# 2

# Understanding Sedimentology

The study of sediments such silt, sand and clay is referred to as sedimentation. It also deals with the study of the processes that result in their formation such as weathering and erosion, transport and deposition. The chapter closely examines the key concepts of sedimentology to provide an extensive understanding of the subject.

Sedimentology encompasses the study of modern sediments such as sand, mud (silt),and clay, and understanding the processes that deposit them. It also compares these observations to studies of ancient sedimentary rocks. Sedimentologists apply their understanding of modern processes to historically formed sedimentary rocks, allowing them to understand how they formed.

Heavy minerals (dark) deposited in quartz beach sand in Chennai, India.

Sedimentary rocks cover most of the Earth's surface, record much of the Earth's history, and harbor the fossil record. Sedimentology is closely linked to stratigraphy, the study of the physical and temporal relationships between rock layers or strata. Sedimentary rocks are useful in various applications, such as for art and architecture, petroleum extraction, ceramicproduction, and checking reservoirs of groundwater.

## Basic Principles

The aim of sedimentology, studying sediments, is to derive information on the depositional conditions that acted to deposit the rock unit, and the relation of the individual rock units in a basin into a coherent understanding of the evolution of the sedimentary sequences and basins, and thus, the Earth's geological history as a whole.

Uniformitarian geology works on the premise that sediments within ancient sedimentary rocks were deposited in the same way as sediments that are being deposited on the Earth's surface today.

In other words, the processes affecting the Earth today are the same as in the past, which then becomes the basis for determining how sedimentary features in the rock record were formed. One may compare similar features today—for example, sand dunes in the Sahara or the Great Sand Dunes National Park near Alamosa, Colorado—to ancient sandstones, such as the Wingate Sandstone of Utah and Arizona, of the southwest United States. Since both have the same features, both can be shown to have formed from aeolian (wind) deposition.

Sedimentological conditions are recorded within the sediments as they are laid down; the form of the sediments at present reflects the events of the past and all events which affect the sediments, from the source of the sedimentary material to the stresses enacted upon them after diagenesis are available for study.

The principle of superposition is critical to the interpretation of sedimentary sequences, and in older metamorphic terrains or fold and thrust beltsm where sediments are often intensely folded or deformed, recognizing younging indicators or fining up sequences is critical to interpretation of the sedimentary section and often the deformation and metamorphic structure of the region.

Folding in sediments is analyzed with the principle of original horizontality, which states that sediments are deposited at their angle of repose which, for most types of sediment, is essentially horizontal. Thus, when the younging direction is known, the rocks can be "unfolded" and interpreted according to the contained sedimentary information.

The principle of lateral continuity states that layers of sediment initially extend laterally in all directions unless obstructed by a physical object or topography.

The principle of cross-cutting relationships states that whatever cuts across or intrudes into the layers of strata is younger than the layers of strata.

## Methodology

The methods employed by sedimentologists to gather data and evidence on the nature and depositional conditions of sedimentary rocks include:

- Measuring and describing the outcrop and distribution of the rock unit:
  - Describing the rock formation, a formal process of documenting thickness, lithology, outcrop, distribution, contact relationships to other formations.
  - Mapping the distribution of the rock unit, or units.
- Descriptions of rock core (drilled and extracted from wells during hydrocarbon exploration).
- Sequence stratigraphy:
  - Describes the progression of rock units within a basin.
- Describing the lithology of the rock:
  - Petrology and petrography; particularly measurement of texture, grain size, grain shape (sphericity, rounding, and so on), sorting and composition of the sediment.

- Analyzing the geochemistry of the rock:

  ◦ Isotope geochemistry, including use of radiometric dating, to determine the age of the rock, and its affinity to source regions.

## Sedimentary Rock Types

Middle Triassic marginal marine sequence of siltstones and sandstones, southwestern Utah.

There are four primary types of sedimentary rocks: Clastics, carbonates, evaporites, and chemical.

- Clastic rocks are composed of particles derived from the weathering and erosion of precursor rocks and consist primarily of fragmental material. Clastic rocks are classified according to their predominant grain size and their composition. In the past, the term "Clastic Sedimentary Rocks" were used to describe silica-rich clastic sedimentary rocks, however there have been cases of clastic carbonate rocks. The more appropriate term is siliciclastic sedimentary rocks.

  ◦ Organic sedimentary rocks are important deposits formed from the accumulation of biological detritus, and form coal and oil shale deposits, and are typically found within basins of clastic sedimentary rocks.

- Carbonates are composed of various carbonate minerals (most often calcium carbonate ($CaCO_3$)) precipitated by a variety of organic and inorganic processes. Typically, most carbonate rocks are composed of reef material.

- Evaporites are formed through the evaporation of water at the Earth's surface and are composed of one or more salt minerals, such as halite or gypsum.

- Chemical sedimentary rocks, including some carbonates, are deposited by precipitation of minerals from aqueous solution. These include jaspilite and chert.

## Importance of Sedimentary Rocks

Sedimentary rocks provide a multitude of products that both ancient and modern societies have come to utilize.

- Art: Marble, although a metamorphosed limestone, is an example of the use of sedimentary rocks in the pursuit of aesthetics and art.

- Architectural uses: Stone derived from sedimentary rocks is used for dimension stone and in architecture, notably slate, a meta-shale, for roofing, sandstone for load-bearing buttresses.

- Ceramics and industrial materials: Clay for pottery and ceramics including bricks; cement and lime derived from limestone.

- Economic geology: Sedimentary rocks host large deposits of SEDEX ore deposits of lead-zinc-silver, large deposits of copper, deposits of gold, tungsten, and many other precious minerals, gemstones, and industrial minerals including heavy mineral sands ore deposits.

- Energy: Petroleum geology relies on the capacity of sedimentary rocks to generate deposits of petroleum oils. Coal and oil shale are found in sedimentary rocks. A large proportion of the world's uranium energy resources are hosted within sedimentary successions.

- Groundwater: Sedimentary rocks contain a large proportion of the Earth's groundwater aquifers. Human understanding of the extent of these aquifers and how much water can be withdrawn from them depends critically on knowledge of the rocks that hold them (the reservoir).

# SEDIMENTARY FACIES

Sedimentary facies studies the physical, chemical, and biological aspects of a sedimentary bed and the lateral change within sequences of beds of the same geologic age. Sedimentary rocks can be formed only where sediments are deposited long enough to become compacted and cemented into hard beds or strata. Sedimentation commonly occurs in areas where the sediment lies undisturbed for many years in sedimentary basins. Whereas some such basins are small, others occupy thousands of square kilometres and usually have within them several different local depositional environments. Physical, chemical, and biological factors influence these environments, and the conditions that they produce largely determine the nature of the sediments that accumulate. Several different local (sedimentary) environments may thus exist side by side within a basin as conditions change laterally; the sedimentary rocks that ultimately are produced there can be related to these depositional environments. These different but contemporaneous and juxtaposed sedimentary rocks are known as sedimentary facies, a term that was first used by the Swiss geologist Amanz Gressly in 1838.

Sedimentary facies are either terrigenous, resulting from the accumulation of particles eroded from older rocks and transported to the depositional site; biogenic, representing accumulations of whole or fragmented shells and other hard parts of organisms; or chemical, representing inorganic precipitation of material from solution. As conditions change with time, so different depositional sites may change their shapes and characteristics. Each facies thus has a three-dimensional configuration and may in time shift its position.

There are several ways of describing or designating sedimentary facies. By noting the prime physical (or lithological) characteristics, one is able to recognize lithofacies. The biological (or more correctly, paleontological) attributes—the fossils—define biofacies. Both are the direct result of the depositional history of the basin. By ascribing modes of origin to different facies (*i.e.*, interpreting the lithofacies or biofacies) one can visualize a genetic system of facies. It is also common to speak of alluvial facies, bar facies, or reef facies, using the environment as a criterion. This may lead to confusion when revisions of interpretation have to be made because of new or more accurate information about the rocks themselves.

Just as there are regular associations of different local environments in modern sedimentary basins, associations of facies also are known to follow similar patterns in the stratigraphic column. A common example of the latter is that of regular lithofacies and biofacies successions being formed between the edge, or shoreline, of a water-filled basin and the deeper water at its middle. Coarse sediment gives way to finer sediment in the deepening water. Changes in sea level as time passes are a common cause of successive changes in the stratigraphic column. As sea level rises and the sea spreads across what was land, shallow-water sediments are laid down in the newest area to receive such material while areas that were shallow are now deeper and receive finer, or otherwise different, sediments. As the sea advances inland, the belts of sedimentation follow and the retreat of the sea causes the belts to move back offshore.

Johannes Walther, a German geologist, noted in 1894 that the vertical facies sequence in a sedimentary basin undergoing expansion and deepening so that the sea transgresses the land surface (or the reverse, a regression) is the same as the horizontal sequence. This has enabled geologists, knowing the pattern of facies at the surface, to predict accurately what may also be found at depth within a sedimentary basin. It is clear, however, that Walther's observation only applies where there is no major break (*i.e.*, an erosional interval) in the continuity of the succession.

From studies of facies relationships to one another it has become recognized that the gradational, sharp, or eroded contacts between these rock bodies are also of significance in finding the mode of origin. It is also apparent that many facies follow one another in time and space in a repetitive way. A vertical pattern, for example, may be found in a borehole sunk vertically through a sequence of facies. This has been observed in many alluvial sequences and in the coal-bearing series of Carboniferous, Permian, and other systems. Facies under clay, coal, shale, and sandstone may be repeated many times and are called cyclothems. Cyclic or rhythmic sedimentation has been recorded in different rocks in many parts of the world and may arise in many ways; however, re-examination of many successions originally described as cyclic shows that this phenomenon is not as common or as constant as had been believed.

Today it is recognized that facies associations and distribution depend upon interrelated controls. The most important include sedimentary processes, sediment supply, climate, tectonics (earth movements), sea level changes, biological activity, water chemistry, and volcanic activity. Of these the environment of deposition (climate) and tectonic activity are paramount as they may ultimately regulate the other factors.

In industries that exploit earth resources such as fossil fuels, facies (or sedimentary basin) analysis is important in research. It may lead to predictions about where coal, petroleum, natural gas, or other sedimentary materials may be found. Apart from examination of rock specimens, this kind of analysis may also rely heavily upon the geophysical properties of the rocks, such as their densities and electrical magnetic and radioactive properties.

# STRATIGRAPHY

Stratigraphy is a scientific discipline concerned with the description of rock successions and their interpretation in terms of a general time scale. It provides a basis for historical geology, and its principles and methods have found application in such fields as petroleum geology and archaeology.

Stratigraphic studies deal primarily with sedimentary rocks but may also encompass layered igneous rocks (e.g., those resulting from successive lava flows) or metamorphic rocks formed either from such extrusive igneous material or from sedimentary rocks.

A common goal of stratigraphic studies is the subdivision of a sequence of rock strata into mappable units, determining the time relationships that are involved, and correlating units of the sequence—or the entire sequence—with rock strata elsewhere. Following the failed attempts during the last half of the 19th century of the International Geological Congress (IGC; founded 1878) to standardize a stratigraphic scale, the International Union of Geological Sciences (IUGS; founded 1961) established a Commission on Stratigraphy to work toward that end. Traditional stratigraphic schemes rely on two scales: (1) a time scale (using eons, eras, periods, epochs, ages, and chrons), for which each unit is defined by its beginning and ending points, and (2) a correlated scale of rock sequences (using systems, series, stages, and chronozones). These schemes, when used in conjunction with other dating methods—such as radiometric dating (the measurement of radioactive decay), paleoclimatic dating, and paleomagnetic determinations—that, in general, were developed within the last half of the 20th century, have led to somewhat less confusion of nomenclature and to ever more reliable information on which to base conclusions about Earth history.

Because oil and natural gas almost always occur in stratified sedimentary rocks, the process of locating petroleum reservoir traps has been facilitated significantly by the use of stratigraphic concepts and data.

An important principle in the application of stratigraphy to archaeology is the law of superposition—the principle that in any undisturbed deposit the oldest layers are normally located at the lowest level. Accordingly, it is presumed that the remains of each succeeding generation are left on the debris of the last.

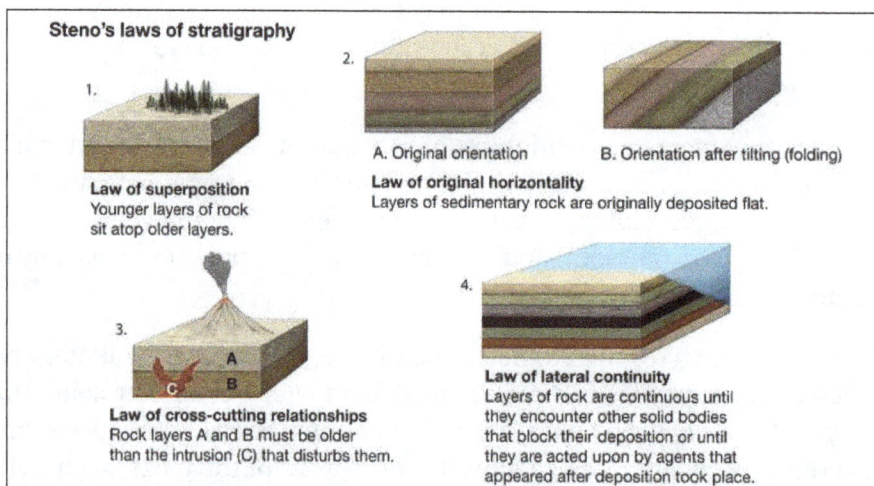

Steno's four laws of stratigraphy.

## Biostratigraphy

Biostratigraphy is the branch of stratigraphy which focuses on correlating and assigning relative ages of rock strata by using the fossil assemblages contained within them. Usually the aim is correlation, demonstrating that a particular horizon in one geological section represents the same period of time as another horizon at some other section. The fossils are useful because sediments of the same age can look completely different because of local variations in the sedimentary environment. For example, one section might have been made up of clays and marls while another has more chalky limestones, but if the fossil species recorded are similar, the two sediments are likely to have been laid down at the same time.

Biostratigraphy originated in the early 19th century, where geologists recognised that the correlation of fossil assemblages between rocks of similar type but different age decreased as the difference in age increased. The method was well-established before Charles Darwin explained the mechanism behind it—evolution.

The first reef builder is a worldwide index fossil for the Lower Cambrian.

Ammonites, graptolites, archeocyathids, and trilobites are index fossils that are widely used in biostratigraphy. Microfossils such as acritarchs, chitinozoans, conodonts, dinoflagellate cysts, ostracods, pollen, spores and foraminiferans are also frequently used. Different fossils work well for sediments of different ages; trilobites, for example, are particularly useful for sediments of Cambrian age. To work well, the fossils used must be widespread geographically, so that they can occur in many different places. They must also be short lived as a species, so that the period of time during which they could be incorporated in the sediment is relatively narrow. The longer lived the species, the poorer the stratigraphic precision, so fossils that evolve rapidly, such as ammonites, are favoured over forms that evolve much more slowly, like nautiloids. Often biostratigraphic correlations are based on a fauna, not an individual species, as this allows greater precision. Further, if only one species is present in a sample, it can mean that (1) the strata were formed in the known fossil range of that organism; (2) that the fossil range of the organism was incompletely known, and the strata extend the known fossil range. For instance, the presence of the trace fossil Treptichnus pedum was used to define the base of the Cambrian period, but it has since been found in older strata.

Fossil assemblages were traditionally used to designate the duration of periods. Since a large change in fauna was required to make early stratigraphers create a new period, most of the periods we recognise today are terminated by a major extinction event or faunal turnover.

## Fossils as a basis for Stratigraphic Subdivision

## Concept of Stage

A stage is a major subdivision of strata, each systematically following the other each bearing a unique assemblage of fossils. Therefore, stages can be defined as a group of strata containing the same major fossil assemblages. French palaeontologist Alcide d'Orbigny is credited for the invention of this concept. He named stages after geographic localities with particularly good sections of rock strata that bear the characteristic fossils on which the stages are based.

## Concept of Zone

In 1856 German palaeontologist Albert Oppel introduced the concept of zone (also known as biozones or Oppel zone). A zone includes strata characterised by the overlapping range of fossils. They represent the time between the appearance of species chosen at the base of the zone and the appearance of other species chosen at the base of the next succeeding zone. Oppel's zones are named after a particular distinctive fossil species, called an index fossil. Index fossils are one of the species from the assemblage of species that characterise the zone.

The zone is the fundamental biostratigraphic unit. Its thickness range from a few to hundreds of metres, and its extant range from local to worldwide. Biostratigraphic units are divided into six principal kinds of biozones:

- Taxon range biozones represent the known stratigraphic and geographic range of occurrence of a single taxon.

- Concurrent range biozone include the concurrent, coincident, or overlapping part of the range of two specified taxa.

- Interval biozone include the strata between two specific biostratigraphic surfaces. It can be based on lowest or highest occurrences.

- Lineage biozone are strata containing species representing a specific segment of an evolutionary lineage.

- Assemblage biozones are strata that contain a unique association of three or more taxa.

- Abundance biozones are strata in which the abundance of a particular taxon or group of taxa is significantly greater than in the adjacent part of the section.

## Index Fossils

Amplexograptus, a graptolite index fossil, from the Ordovician near Caney Springs, Tennessee.

To be useful in stratigraphic correlation index fossils should be:

- Independent of their environment.

- Geographically widespread (provincialism/isolation of species should be avoided as much as possible).

- Rapidly evolving.

- Easy to preserve (Easier in low-energy, non-oxidized environment).

- Easy to identify.

## Faunal Succession

Fossil organisms succeed one another in a definite and determinable order and therefore any time period can be recognized by its fossil content.

## Chronostratigraphy

Chronostratigraphy is the branch of stratigraphy that studies the age of rock strata in relation to time. The ultimate aim of chronostratigraphy is to arrange the sequence of deposition and the time of deposition of all rocks within a geological region, and eventually, the entire geologic record of the Earth.

The standard stratigraphic nomenclature is a chronostratigraphic system based on palaeonto-logical intervals of time defined by recognised fossil assemblages (biostratigraphy). The aim of chronostratigraphy is to give a meaningful age date to these fossil assemblage intervals and interfaces.

## Methodology

Chronostratigraphy relies heavily upon isotope geology and geochronology to derive hard dating of known and well defined rock units which contain the specific fossil assemblages defined by the stratigraphic system. As it is practically very difficult to isotopically date most fossils and sedimentary rocks directly, inferences must be made in order to arrive at an age date which reflects the beginning of the interval.

The methodology used is derived from the law of superposition and the principles of cross-cutting relationships. Because igneous rocks occur at specific intervals in time and are essentially instantaneous on a geologic time scale, and because they contain mineral assemblages which may be dated more accurately and precisely by isotopic methods, the construction of a chronostratigraphic column relies heavily upon intrusive and extrusive igneous rocks.

Metamorphism, often associated with faulting, may also be used to bracket depositional intervals in a chronostratigraphic column. Metamorphic rocks can occasionally be dated, and this may give some limits to the age at which a bed could have been laid down. For example, if a bed containing graptolites overlies crystalline basement at some point, dating the crystalline basement will give a maximum age of that fossil assemblage.

This process requires a considerable degree of effort and checking of field relationships and age dates. For instance, there may be many millions of years between a bed being laid down and an intrusive rock cutting it; the estimate of age must necessarily be between the oldest cross-cutting intrusive rock in the fossil assemblage and the youngest rock upon which the fossil assemblage rests.

## Units

Chronostratigraphic units, with examples:

- Eonothem – Phanerozoic.

- Erathem – Paleozoic.

- System – Ordovician.

- Series – Upper Ordovician.

- Stage – Ashgill.

## Differences from Geochronology

It is important not to confuse geochronologic and chronostratigraphic units. Chronostratigraphic units are geological material, so it is correct to say that fossils of the species *Tyrannosaurus rex* have been found in the Upper Cretaceous Series. Geochronological units are periods of time and take the same name as standard stratigraphic units but replacing the terms upper/lower with late/early. Thus it is also correct to say that *Tyrannosaurus rex* lived during the Late Cretaceous Epoch.

Chronostratigraphy is an important branch of stratigraphy because the age correlations derived are crucial to drawing accurate cross sections of the spatial organization of rocks and to preparing accurate paleogeographic reconstructions.

## Lithostratigraphy

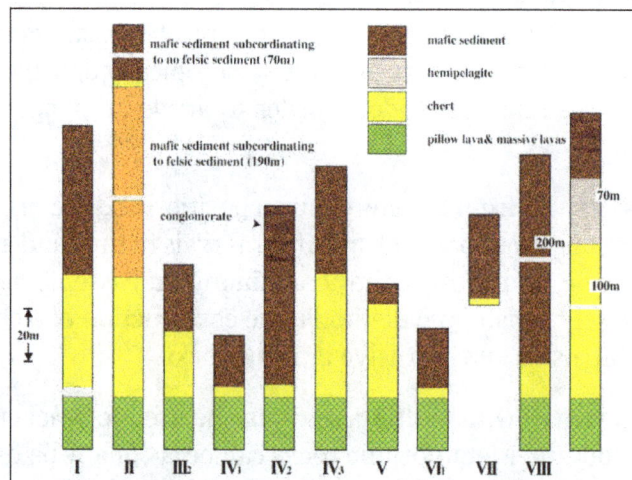

In lithostratigraphy rock units are considered in terms of the lithological characteristics of the strata and their relative stratigraphic positions. The relative stratigraphic positions of rock units can be determined by considering geometric and physical relationships that indicate which beds are older and which ones are younger. The units can be classified into a hierarchical system of members, formations and groups that provide a basis for categorising and describing rocks in lithostratigraphic terms.

## Stratigraphic Relationships

## Superposition

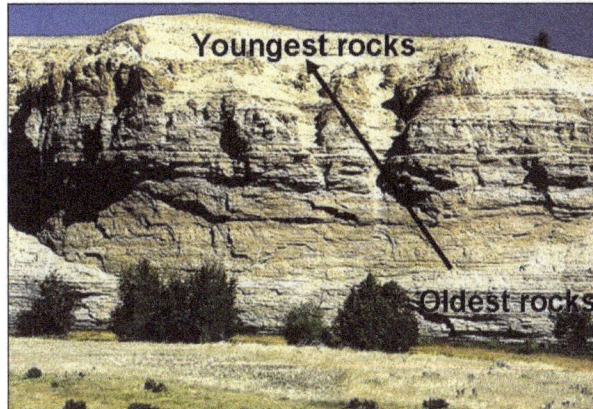

Principle of superposition.

Provided the rocks are the right way up the beds higher in the stratigraphic sequence of deposits will be younger than the lower beds. This rule can be simply applied to a layer-cake stratigraphy but must be applied with care in circumstances where there is a significant depositional topography (e.g. fore-reef deposits may be lower than reef-crest rocks).

## Unconformities

An unconformity is a break in sedimentation and where there is erosion of the underlying strata this provides a clear relationship in which the beds below the unconformity are clearly older than those above it. All rocks which lie above the unconformity, or a surface that can be correlated with it, must be younger than those below. In cases where strata have been deformed and partly eroded prior to deposition of the younger beds, an angular unconformity is formed. A disconformity marks a break in sedimentation and some erosion, but without any deformation of the underlying strata.

## Cross-Cutting Relationships

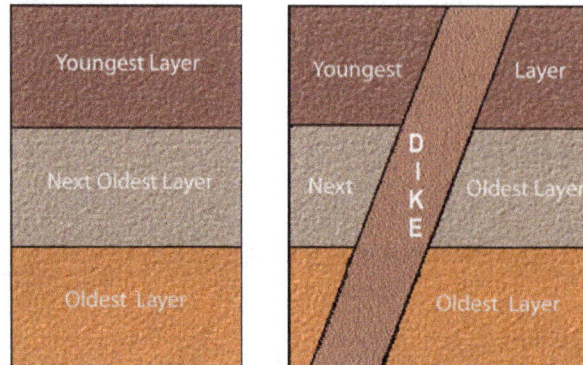

Any unit that has boundaries that cut across other strata must be younger than the rocks it cuts. This is most commonly seen with intrusive bodies such as batholiths on a larger scale and dykes on a smaller scale. This relationship is also seen in fissure fills, sedimentary dykes that form by younger sediments filling a crack or chasm in older rocks.

## Included Fragments

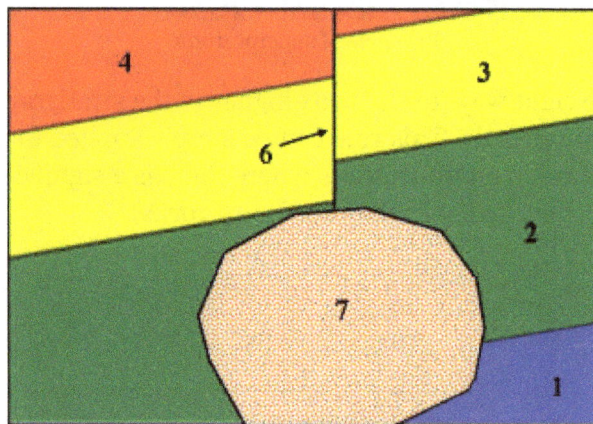

The fragments in a clastic rock must be made up of a rock that is older than the strata in which they are found. The same relationship holds true for igneous rocks that contain pieces of the surrounding country rock as xenoliths (literally 'foreign rocks'). This relationship can be useful in determining the age relationship between rock units that are some distance apart. Pebbles of a characteristic lithology can provide conclusive evidence that the source rock type was being eroded by the time a later unit was being deposited tens or hundreds of kilometres away.

## Way-up Indicators in Sedimentary Rocks

The folding and faulting of strata during mountain building can rotate whole successions of beds (formed as horizontal or nearly horizontal layers) through any angle, resulting in beds that may be vertical or completely overturned. In any analysis of deformed strata, it is essential to know the direction of younging, that is, the direction through the layers towards younger rocks. The direction of younging can be determined by small-scale features that indicate the way-up of the beds or by using other stratigraphic techniques to determine the order of formation.

## Lithostratigraphic Units

There is a hierarchical framework of terms used for lithostratigraphic units, and from largest to smallest these are: 'Supergroup', 'Group', 'Formation', 'Member' and 'Bed'. The basic unit of lithostratigraphic division of rocks is the formation, which is a body of material that can be identified by its lithological characteristics and by its stratigraphic position. It must be traceable laterally, that is, it must be mappable at the surface or in the subsurface. A formation should have some degree of lithological homogeneity and its defining characteristics may include mineralogical composition, texture, primary sedimentary structures and fossil content in addition to the lithological composition. Note that the material does not necessarily have to be lithified and that all the discussion of terminology and stratigraphic relationships applies equally to unconsolidated sediment. A formation is not defined in terms of its age either by isotopic dating or in terms of biostratigraphy. Information about the fossil content of a mapping unit is useful in the description of a formation but the detailed taxonomy of the fossils that may define the relative age in biostratigraphic terms does not form part of the definition of a lithostratigraphic unit. A formation may be, and often is, a diachronous unit, that is, a deposit with the same lithological properties that was formed at different times in different places. A formation may be divided into smaller units in order to provide more detail of the distribution of lithologies. The term member is used for rock units that have limited lateral extent and are consistently related to a particular formation (or, rarely, more than one formation). An example would be a formation composed mainly of sandstone but which included beds of conglomerate in some parts of the area of outcrop. A number of members may be defined within a formation (or none at all) and the formation does not have to be completely subdivided in this way: some parts of a formation may not have a member status. Individual beds or sets of beds may be named if they are very distinctive by virtue of their lithology or fossil content. These beds may have economic significance or be useful in correlation because of their easily recognisable characteristics across an area. Where two or more formations are found associated with each other and share certain characteristics they are considered to form a group. Groups are commonly bound by unconformities which can be traced basin-wide. Unconformities that can be identified as major divisions in the stratigraphy over the area of a continent are sometimes considered to be the bounding surfaces of associations of two or more groups known as a supergroup.

## Description of Lithostratigraphic Units

The formation is the fundamental lithostratigraphic unit and it is usual to follow a certain procedure in geological literature when describing a formation to ensure that most of the following issues are considered. Members and groups are usually described in a similar way.

## Lithology and Characteristics

The field characteristics of the rock, for example, an oolitic grainstone, interbedded coarse siltstone and claystone, a basaltic lithic tuff, and so on form the first part of the description. Although a formation will normally consist mainly of one lithology, combinations of two or more lithologies will often constitute a formation as interbedded or interfingering units. Sedimentary structures (ripple cross-laminations, normal grading, etc.), petrography (often determined from thin-section analysis) and fossil content (both body and trace fossils) should also be noted.

## Definition of Top and Base

These are the criteria that are used to distinguish beds of this unit from those of underlying and overlying units; this is most commonly a change in lithology from, say, calcareous mudstone to coral boundstone. Where the boundary is not a sharp change from one formation to another, but is gradational, an arbitrary boundary must be placed within the transition. As an example, if the lower formation consists of mainly mudstone with thin sandstone beds, and the upper is mainly sandstone with subordinate mudstone, the boundary may be placed at the point where sandstone first makes up more than 50% of beds. A common convention is for only the base of a unit to be defined at the type section: the top is taken as the defined position of the base of the overlying unit. This convention is used because at another location there may be beds at the top of the lower unit that are not present at the type locality: these can be simply added to the top without a need for redefining the formation boundaries.

## Type Section

A type section is the location where the lithological characteristics are clear and, if possible, where the lower and upper boundaries of the formation can be seen. Sometimes it is necessary for a type section to be composite within a type area, with different sections described from different parts of the area. The type section will normally be presented as a graphic sedimentary log and this will form the strato type. It must be precisely located (grid reference and/or GPS location) to make it possible for any other geologist to visit the type section and see the boundaries and the lithological characteristics described.

## Thickness and Extent

The thickness is measured in the type section, but variations in the thickness seen at other localities are also noted. The limits of the geographical area over which the unit is recognised should also be determined. There are no formal upper or lower limits to thickness and extent of rock units defined as a formation (or a member or group). The variability of rock types within an area will be the main constraint on the number and thickness of lithostratigraphic units that can be described and defined. Quality and quantity of exposure also play a role, as finer subdivision is possible in areas of good exposure.

## Other Information

Where the age for the formation can be determined by fossil content, radiometric dating or relationships with other rock units this may be included, but note that this does not form part of the definition of the formation. A formation would not be defined as, for example, 'rocks of Burdigalian age', because an interpretation of the fossil content or isotopic dating information is required to determine the age. Information about the facies and interpretation of the environment of deposition might be included but a formation should not be defined in terms of depositional environment, for example, 'lagoonal deposits', as this is an interpretation of the lithological characteristics. It is also useful to comment on the terminology and definitions.

## Lithostratigraphic Nomenclature

It helps to avoid confusion if the definition and naming of stratigraphic units follows a set of rules. Formal codes have been set out in publications such as the 'North American Stratigraphic Code' and the

'International Stratigraphic Guide'. A useful summary of stratigraphic methods, which is rather more user-friendly than the formal documents, is a handbook called 'Stratigraphical Procedure'. The name of the formation, group or member must be taken from a distinct and permanent geographical feature as close as possible to the type section. The lithology is often added to give a complete name such as the Kingston Limestone Formation, but it is not essential, or necessarily desirable if the lithological characteristics are varied. The choice of geographical name should be a feature or place marked on topographic maps such as a river, hill, town or village. The rules for naming members, groups and super groups are essentially the same as for formations, but note that it is not permissible to use a name that is already in use or to use the same name for two different ranks of lithostratigraphic unit. There are some exceptions to these rules of nomenclature that result from historical precedents, and it is less confusing to leave a well established name as it is rather than to dogmatically revise it. Revisions to stratigraphic nomenclature may become necessary when more detailed work is carried out or more information becomes available. New work in an area may allow a formation to be subdivided and the formation may then be elevated to the rank of group and members may become formations in their own right. For the sake of consistency the geographical name is retained when the rank of the unit is changed.

### Lithodemic Units: Non-Stratiform Rock Units

The concepts of division into stratigraphic units were developed for rock bodies that are stratiform, layered units, but many metamorphic, igneous plutonic and structurally deformed rocks are not stratiform and they do not follow the rules of superposition. Nonstratiform bodies of rock are called lithodemic units. The basic unit is the lithodeme and this is equivalent in rank to a formation and is also defined on lithological criteria. The word 'lithodeme' is itself rarely used in the name: the body of rock is normally referred to by its geographical name and lithology, such as the White River Granite or Black Hill Schist. An association of lithodemes that share lithological properties, such as a similar metamorphic grade, is referred to as a suite: the term complex is also used as the equivalent to a group for volcanic or tectonically deformed rocks.

# CARBONATE PLATFORM

A carbonate platform is a sedimentary body which possesses topographic relief, and is composed of autochthonic calcareous deposits. Platform growth is mediated by sessile organisms whose skeletons build up the reef or by organisms (usually microbes) which induce carbonate precipitation through their metabolism. Therefore, carbonate platforms can not grow up everywhere: they are not present in places where limiting factors to the life of reef-building organisms exist. Such limiting factors are, among others: light, water temperature, transparency and pH-Value. For example, carbonate sedimentation along the Atlantic South American coasts takes place everywhere but at the mouth of the Amazon River, because of the intense turbidity of the water there. Spectacular examples of present-day carbonate platforms are the Bahama Banks under which the platform is roughly 8 km thick, the Yucatan Peninsula which is up to 2 km thick, the Florida platform, the platform on which the Great Barrier Reef is growing, and the Maldive atolls. All these carbonate platforms and their associated reefs are confined to tropical latitudes. Today's reefs are built mainly by scleractinian corals, but in the distant past other organisms, like archaeocyatha (during the Cambrian) or extinct cnidaria (tabulata and rugosa) were important reef builders.

## Carbonate Precipitation from Seawater

What makes carbonate platform environments different from other depositional environments is that carbonate is a product of precipitation, rather than being a sediment transported from elsewhere, as for sand or gravel. This implies for example that carbonate platforms may grow far from the coastlines of continents, as for the Pacific atolls.

The mineralogic composition of carbonate platforms may be either calcitic or aragonitic. Seawater is oversaturated in carbonate, so under certain conditions $CaCO_3$ precipitation is possible. Carbonate precipitation is thermodynamically favoured at high temperature and low pressure. Three types of carbonate precipitation are possible: biotically controlled, biotically induced and abiotic. Carbonate precipitation is biotically controlled when organisms (such as corals) are present that exploit carbonate dissolved in seawater to build their calcitic or aragonitic skeletons. Thus they may develop hard reef structures. Biotically induced precipitation takes place outside the cell of the organism, thus carbonate is not directly produced by organisms, but precipitates because of their metabolism. Abiotic precipitation involves little or no biological influence.

## Classification

The three types of precipitation (abiotic, biotically induced and biotically controlled) cluster into three "carbonate factories". A carbonate factory is the ensemble of the sedimentary environment, the intervening organisms and the precipitation processes that lead to the formation of a carbonate platform. The differences between three factory is the dominant precipitation pathway and skeletal associations. In contrast, a carbonate platform is a geological structure of parautochotonous carbonate sediments and carbonate rocks, having a morphological relief.

## Platforms Produced by the Tropical Factory

In these carbonate factories, precipitation is biotically controlled, mostly by autotrophic organisms. Organisms that build this kind of platforms are today mostly corals and green algae, that need sunlight for photosynthesis and thus live in the euphotic zone (i.e., shallow water environments in which sunlight penetrates easily). Tropical carbonate factories are only present today in warm and sunlit waters of the tropical-subtropical belt, and they have high carbonate production rates but only in a narrow depth window. The depositional profile of a Tropical factory is called "rimmed" and includes three main parts: a lagoon, a reef and a slope. In the reef, the framework produced by large-sized skeletons, as those of corals, and by encrusting organisms resists wave action and forms a rigid build up that may develop up to sea-level. The presence of a rim produces restricted circulation in the back reef area and a lagoon may develop in which carbonate mud is often produced. When reef accretion reaches the point that the foot of the reef is below wave base, a slope develops: the sediments of the slope derive from the erosion of the margin by waves, storms and gravitational collapses. This process accumulates coral debris in clinoforms. The maximum angle that a slope can achieve is the settlement angle of gravel (30-34°).

## Platforms Produced by the Cool-Water Factory

In these carbonate factories, precipitation is biotically controlled by heterotrophic organisms, sometimes in association with photo-autotrophic organisms such as red algae. The typical skeletal association includes foraminifers, red algae and molluscs. Despite being autotrophic, red algae are mostly associated to heterotrophic carbonate producers, and need less light than green algae. The range of occurrence of cool-water factories extends from the limit of the tropical factory (at about 30°) up to polar latitudes, but they could also occur at low latitudes in the thermocline below the warm surface waters or in upwelling areas. This type of factories has a low potential of carbonate production, is largely independent from sunlight availability, and can sustain a higher amount of nutrients than tropical factories. Carbonate platforms built by the "cool-water factory" show two types of geometry or depositional profile, i.e., the homoclinal ramp or the distally-steepened ramp. In both geometries there are three parts: the inner ramp above the fair weather wave base, the middle ramp, above the storm wave base, the outer ramp, below the storm wave base. In distally steepened ramps, a distal step is formed between the middle and outer ramp, by the in situ accumulation of gravel-sized carbonate grains.

## Platforms Produced by the Mud-Mound Factory

These factories are characterised by abiotic precipitation and biotically induced precipitation. The typical enivronmental settings where "mud-mound factories" are found in the Phanerozoic are dysphotic or aphotic, nutrient-rich waters that are low in oxygen but not anoxic. These conditions often prevail in the thermocline, for example at intermediate water depths below the ocean's mixed layer. The most important component of these platforms is fine-grained carbonate that precipitates in situ (automicrite) by a complex interplay of biotic and abiotic reactions with microbes and decaying organic tissue. Mud-mound factories do not produce a skeletal association, but they have specific facies and microfacies, for example stromatolites, that are laminated microbialites, and thrombolites, that are microbialites characterized by clotted peloidal fabric at the microscopic scale and by dendroid fabric at the hand-sample scale. The geometry of these platforms is mound-shaped, where all the mound is productive, including the slopes.

## Geometry of Carbonate Platforms

Several factors influence the geometry of a carbonate platform, including inherited topography, synsedimentary tectonics, exposition to currents and trade winds. Two main types of carbonate platforms are distinguished on the base of their geographic setting: isolated (as Maldives atolls) or epicontinental (as the Belize reefs or the Florida Keys). However, the one most important factor influencing geometries is perhaps the type of carbonate factory. Depending on the dominant carbonate factory, we can distinguish three types of carbonate platforms: T-type carbonate platforms (produced by "tropical factories"), C-type carbonate platforms (produced by "cool-water factories"), M-type carbonate platforms ("produced by mud-mound factories"). Each of them has its own typical geometry.

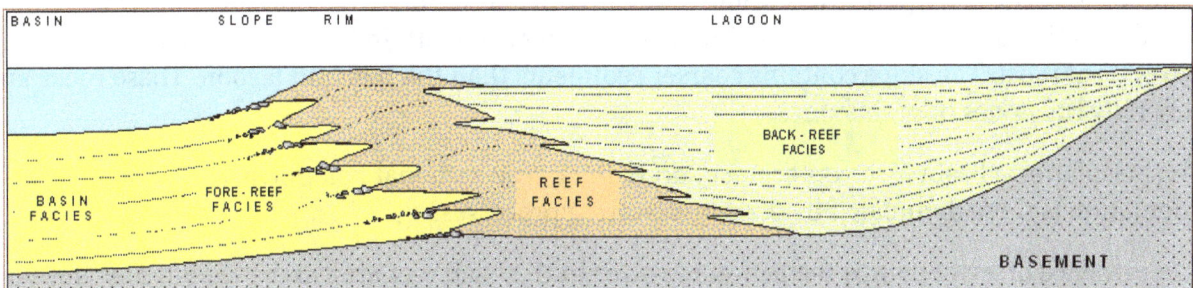

Generalized cross-section of a typical carbonate platform.

## T-type Carbonate Platforms

The depositional profile of T-type carbonate platforms can be subdivided into several sedimentary environments.

The carbonate hinterland is the most landward environment, composed by weathered carbonate rocks. The evaporitic tidal flat is a typical low-energy environment.

An example of carbonate mud sedimentation in the internal part of the Florida Bay lagoon. The presence of young mangroves is important to entrap the carbonate mud.

The internal lagoon, as the name suggests, is the part of platform behind the reef. It is characterised by shallow and calm waters, and so it is a low-energy sedimentary environment. Sediments are composed by reef fragments, hard parts of organisms and, if the platform is epicontinental, also by a terrigenous contribution. In some lagoons (e.g., the Florida Bay), green algae produce great volumes of carbonate mud. Rocks here are mudstones to grainstones, depending on the energy of the environment.

The reef is the rigid structure of carbonate platforms and is located between the internal lagoon and the slope, in the platform margin, in which the framework produced by large-sized skeletons, as those of corals, and by encrusting organisms will resist wave action and form a rigid build up that may develop up to sea-level. Survival of the platform depends on the existence of the reef, because only this part of the platform can build a rigid, wave-resistant structure. The reef is created by essentially in-place, sessile organisms. Today's reefs are mostly built by hermatypic corals. Geologically speaking, reef rocks can be classified as massive boundstones.

The slope is the outer part of the platform, connecting the reef with the basin. This depositional environment acts as sink for excess carbonate sediment: most of the sediment produced in the lagoon and reef is transported by various processes and accumulates in the slope, with an inclination depending on the grain size of sediments, and that could attain the settlement angle of gravel (30-34°) at most. The slope contains coarser sediments than the reef and lagoon. These rocks are generally rudstones or grainstones.

The periplatform basin is the outermost part of the t-type carbonate platform, and carbonate sedimentation is there dominated by density-cascating processes.

The presence of a rim damps the action of waves in the back reef area and a lagoon may develop in which carbonate mud is often produced. When reef accretion reaches the point that the foot of the reef is below wave base, a slope develops: the sediments of the slope derive from the erosion of the margin by waves, storms and gravitational collapses. This process accumulates coral debris in clinoforms. Clinoforms are beds that have a sigmoidal or tabular shape, but are always deposited with a primary inclination.

The size of a T-type carbonate platform, from the hinterland to the foot of the slope, can be of tens of kilometers.

## C-type Carbonate Platforms

C-type carbonate platforms are characterized by the absence of early cementation and lithification, and so the sediment distribution is only driven by waves and, in particular, it occurs above the wave base. They show two types of geometry or depositional profile, i.e., the homoclinal ramp or the distally-steepened ramp. In both geometries there are three parts. In the inner ramp, above the fair weather wave base, the carbonate production is slow enough that all sediments may be transported offshore by waves, currents and storms. As a consequence, the shoreline may be retreating, and so in the inner ramp there may be a cliff caused by erosional processes. In the middle ramp, between the fair weather wave base and the storm wave base, carbonate sediment remains in place and can be reworking only by the storm waves. In the outer ramp, below the storm wave base, fine sediments may accumulate. In distally steepened ramps, a distal step is formed between the middle and outer ramp, by the in situ accumulation of gravel-sized carbonate grains (e.g., rhodoliths) only episodically moved by currents. Carbonate production occurs along the full depositional profile in this type of carbonate platforms, with an extra production in the outer part of the middle ramp, but carbonate production rates are always less than in the T-type carbonate platforms.

## M-type Carbonate Platforms

M-type carbonate platforms are characterized by an inner platform, an outer platform, an upper slope made by microbial boundstone, and a lower slope often made by breccia. The slope may be steeper than the angle of repose of gravels, with an inclination that may attain 50°.

In the M-type carbonate platforms the carbonate production mostly occurs on the upper slope and in the outer part of the inner platform.

The Cimon del Latemar (Trento province, Dolomites, northern Italy) represents the internal lagoon of a fossil carbonate platform. Continuous sedimentation took place in an environment as the one described in the image of the Florida Bay and, given a strong subsidence, led to the formation of a sedimentary series that therefore acquired considerable thickness.

## Carbonate Platforms in the Geological Record

Sedimentary sequences show carbonate platforms as old as the Precambrian, when they were formed by stromatolitic sequences. In the Cambrian carbonate platforms were built by archaeocyatha. During Paleozoic brachiopod (richtofenida) and stromatoporoidea reefs were erected. At the middle of the Paleozoic era corals became important platforms builders, first with tabulata (from the Silurian) and then with rugosa (from the Devonian). Scleractinia become important reef builders beginning only in the Carnian (upper Triassic). Some of the best examples of carbonate platforms are in the Dolomites, deposited during the Triassic. This region of the Southern Alps contains many well preserved isolated carbonate platforms, including the Sella, Gardenaccia, Sassolungo and Latemar. The middle Liassic "bahamian type" carbonate platform of Morocco is characterised by the accumulation of autocyclic regressive cycles, spectacular supratidal deposits and vadose diagenetic features with dinosaur tracks. The Tunisian coastal "chotts" and their cyclic muddy deposits represent a good recent equivalent. Such cycles were also observed on the Mesozoic Arabic platform, Oman and Abu Dhabi with the same microfauna of foraminifera in an almost identical biostratigraphic succession.

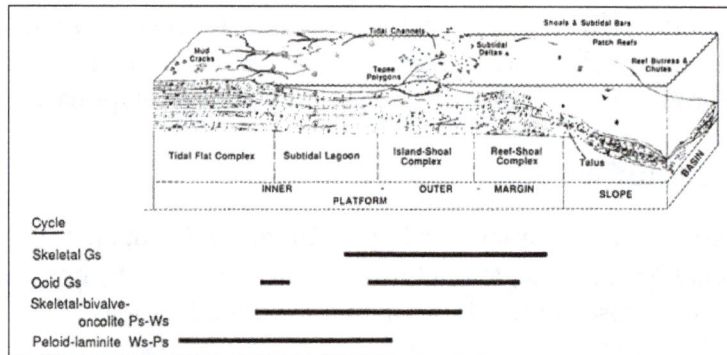

High Atlas middle Liassic carbonate platform of Morocco with first order autocyclic regressive cycles.

Carbonate platform environment.

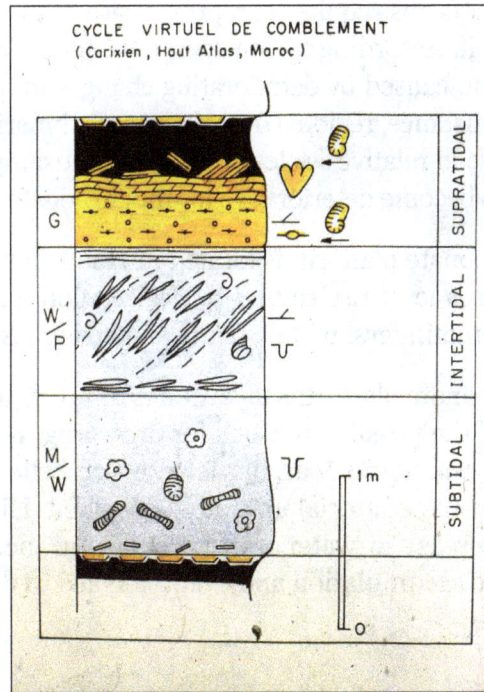

Virtual metric "shallowing upward sequence" observed all along (more than 10,000 km) the south Tethyan margin during middle Liassic times. The (micro)fossils are identical till Oman and beyond. In the Cretaceous period there were platforms built by bivalvia (rudists).

## Sequence Stratigraphy of Carbonate Platforms

With respect to the sequence stratigraphy of siliciclastic systems, carbonate platforms present some peculiarities, which are related to the fact that carbonate sediment is precipitated directly on the platform, mostly with the intervention of living organisms, instead of being only transported and deposited. Among these peculiarities, carbonate platforms may be subject to drowning, and may be the source of sediment via highstand shedding or slope shedding.

## Drowning

Drowning of a carbonate platform is an event where the relative sea level rise is faster than the accumulation rate on a carbonate platform, which eventually leads to the platform to submerge below the euphotic zone. In the geologic record of a drowned carbonate platform, neritic deposits change rapidly into deep-marine sediments. Typically hardgrounds with ferromanganese oxides, phosphate or glauconite crusts lie in between of neritic and deep-marine sediments.

Several drowned carbonate platforms have been found in the geologic record. However, it has not been very clear how the drowning of carbonate platforms exactly happen. Modern carbonate platforms and reefs are estimated to grow approximately 1 000 µm/yr, possibly several times faster in the past. 1 000 µm/yr growth rate of carbonates exceeds by orders of magnitude any relative sea level rise that is caused by long-term subsidence, or changes in eustatic sea level. Based on the rates of these processes, drowning of the carbonate platforms should not be possible, which causes "the paradox of drowned carbonate platforms and reefs".

Since drowning of carbonate platforms requires exceptional rise in the relative sea level, only limited number of processes can cause it. According to Schlager, only anomalously quick rise of relative sea level or benthic growth reduction caused by deteriorating changes in the environment could explain the drowning of platforms. For instance, regional downfaulting, submarine volcanism or glacioeustacy could be the reason for rapid rise in relative sea level, whereas for example changes in oceanic salinity might cause the environment to become deteriorative for the carbonate producers.

One example of a drowned carbonate platform is located in Huon Gulf, Papua New Guinea. It is believed to be drowned by rapid sea level rise caused by deglaciation and subsidence of the platform, which enabled coralline algal-foraminiferal nodules and halimeda limestones to cover the coral reefs.

Plate movements carrying carbonate platforms to latitudes unfavourable for carbonate production are also suggested to be one of the possible reasons for drowning. For example, guyots located in the Pacific Basin between Hawaiian and Mariana Islands are believed to be transported to low southern latitudes (0-10°S) where equatorial upwelling occurred. High amounts of nutrients and higher productivity caused decrease in water transparency and increase in bio-eroders populations, which reduced carbonate accumulation and eventually led to drowning.

## Highstand Shedding

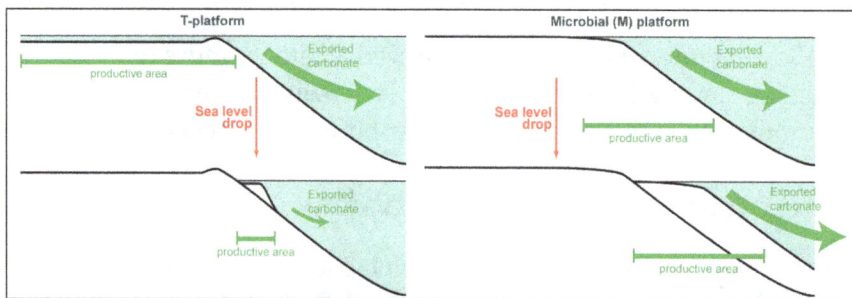

Highstand shedding and slope shedding.

Highstand shedding is a process in which a carbonate platform produces and sheds most of the sediments into the adjacent basin during highstands of sea level. This process has been observed on all rimmed carbonate platforms in the Quaternary, such as the Great Bahama Bank. Flat topped, rimmed platforms with steep slopes show more pronounced highstand shedding than platforms with gentle slopes and cool water carbonate systems.

Highstand shedding is pronounced on tropical carbonate platforms because of the combined effect of sediment production and diagenesis. Sediment production of a platform increases with its size, and during highstand the top of the platform is flooded and the productive area is bigger compared to the lowstand conditions, when only a minimal part of the platform is available for production. The effect of increased highstand production is enhanced by the rapid lithification of carbonate during lowstands, because the exposed platform top is karstified rather than eroded, and does not export sediment.

## Slope Shedding

Slope shedding is a process typical of microbial platforms, in which the carbonate production is nearly independent from sea level oscillations. The carbonate factory, composed of microbial communities precipitating microbialites, is insensitive to light and can extend from the platform break down the slope to hundreds of meters in depth. Sea level drops of any reasonable amplitude would not significantly affect the slope production areas. Microbial boundstone slope systems are remarkably different from tropical platforms in sediment productions profiles, slope readjustment processes and sediment sourcing. Their progradation is independent from platform sediment shedding and largely driven by slope shedding.

Examples of margins that may be affected of slope shedding that are characterized by various contributions of microbial carbonate growth to the upper slope and margin, are:

- The Canning Basin in Australia.

- The Guilin platform in the southern China.

- The Permian of the US Permian Basin.

- The middle Triassic carbonate platforms of the Dolomites.

# SEDIMENTARY ENVIRONMENTS

Any area where sediment is capable of accumulating is considered a depositional environment. Each depositional environment posseses distinctive physical, chemical, and biological characteristics that allow for specific kinds of deposits. how the properties of individual environments, found in continental, transitional, and marine settings, influence deposition and how this information is used to infer paleoenvironments.

Continental enviroments refers to all depostion that occurs on land; this includes fluvial systems (rivers and streams), lakes, deserts and areas adjacent to or covered by glaciers. At first glance it may seem impossible to be able to determine exactly which environment produced a given sediment but the factors that allow deposition behave extremely different in each environment.

Braided Stream, Alaska.

Fluvial environments are those environments that are dominated by runnin water and are characterized as either a meandering stream system or a braided stream system. Braided stream systems are typically found in areas with high slopes or in areas where the water contains a high sediment load. In either case, braided stream systems are characterized by fast moving water in multiple shallow channels that appear to be braided. These channels form when fast moving water dramatically slows and the larger particles (such as gravels) are deposited within the channel and block the flow of water. Because braided stream systems are high energy smaller clast sizes, such as silts and clays, are not deposited instead they are washed further down stream. As a result braided stream systems can be identified by their shallow channels of crossbedded-sands, gravel deposits, and the lack of mud or clays.

The Mississippi River, example of a meandering stream.

Unlike braided stream systems, meandering stream systems have a single stream channel and are found on gently sloped or flat areas. As a result of the gentle slopes water moves relatively slowly and is therefore only capable of carrying sediments no larger than sand. As the meandering stream moves across the land point bars develop on the inside curve of each turn. Point bars are depositional features that consist of crossbedded-sands and caused by the change in velocity of water as it moves through a curved channel. Meandering streams also develop large flood plains on either side of the stream channel. During flood stages excess water carrying suspended particles of silts and clays flows away from the fast-moving channel water. As a result this water is slowed and the finer particles deposited on the flood plain.

Alluvial Fan forming at the mouth of a canyon.

Because of their size desert environments are typically identified by the assoication of many geologic features, such as sand dunes (Large scale cross-bedded sands), alluvial fan deposits, and playa lake deposits. Alluvial fans form where streams and debris flows discharge from mountains onto a valley floor leaving a triagular, or wedge-shaped, deposit of sands and gravel; typically the larger clasts are more abundant closer to their sources. Beyond the alluvial fans wind flowing down the mountain slope is able to pick up previously deposited sands and redeposit them along the desert floor in sand dunes. Lastly, many deserts contain lakes formed from seasonal rains called playa lakes. Because the water contained within a lake is extremely still it is a low energy environ-ment. During rainy seasons the water washing into the basin of the lake will carry the sands, silts, and clays off the dry desert floor. As this water enters the newly formed lake the velocity of the water drops to nearly zero and throughout the dry season the finer particles of clay are capable of settling on the lake floor. Because this process occurrs every year playa lake deposits produce thin layers of mud, called varves, that record the seasonal deposition.

The sediment deposited in glacial environments is collectively known as drift. Glacial drift is deposited through a number of processes and is categorized as either till or outwash. Till is deposited directly by the melting ice along the glacial margin and consists of all sediment that has been picked up by the glacier as it scraped across the surface of the Earth. Unlike fluvial systems where the speed of the liquid water controls the maximum size particle that can be transported, glaciers have no limit to the size clast they are capable of transporting. Although they may move extremely slow-- maybe only a few inches a day-- any object that is incorporated into the icy structure is simply moved along with the ice and is eventually deposited in linear piles along the glacial margin. These linear piles of till are known as gla-cial moraines. Outwash deposits form as meltwater from the glacier deposits sands and silts in braided stream environments directly in front of the glacial margin.

Transitional environments are found anywhere both marine and land processes are responsible for deposition. These environments include beaches, deltas, barrier islands, and tidal flats. Deltas are accumulations of sediment that form when a stream enters a relatively quiet (low energy) body of wa-ter such as a bay, sea, or lake. Deltas get their name from the triangular shape that is often produced by this type of deposition. The true shape of a delta is controlled by both deposition and the rate of erosion. In areas where the rate of deposition is greater than erosion a delta will continually prograde into quieter body of water, such as the case with the Mississippi Delta. Where the rate of deposition is similar to the rate of erosion the delta will develop the stereotypical triangular shape like the Nile River Delta.

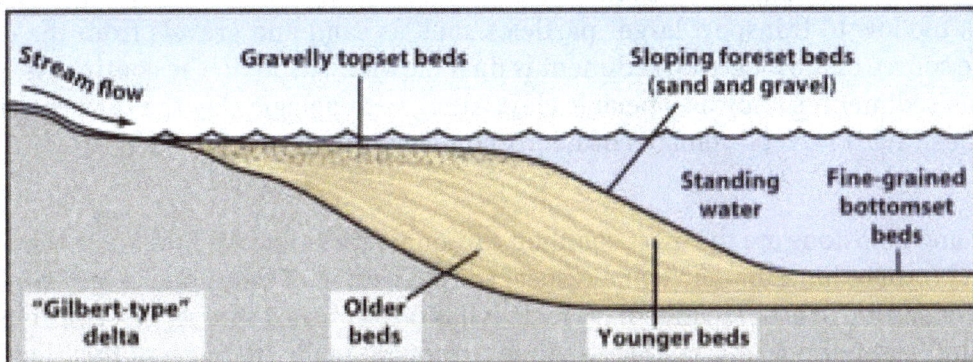

Formation of a Gilbert-type delta.

Regardless of the type of delta produced sand deposited within ancient deltas can be identified by the presence of foreset and bottomset beds. Foreset beds form when sediment laden fast moving water enter the low energy quiet water environment. When the velocity of the water drops the heavier sand sized particles are deposited at an incline along the floor of the open water body. Subsequent deposition occurs on top of the previously deposited foreset beds causing the shoreline to prograde into the open water environment. Fine-grained material deposited transported by the river system is deposited away from shore in the lower energy environment and deposited nearly horizontal in what are know as bottomset beds.

Barrier islands lie offshore and are separated from the mainland by a lagoon or bay. On Long Island barrier island is known as Fire Island and the lagoon is the Great South Bay. Barrier Islands are typically composed of sands which grade into finer deposits offshore. Cross-bedded dunes are also common features of barrier islands.

## Marine Environments

Marine environments include the continental shelf, slope and rise, and the deep sea floor. Although much of the material eroded from the continents is eventually deposited within the marine environments other material such as those precipitated directly from the water are also deposited. The specific type of material that is deposited in a marine environment is determined by the environment's distance from shore, biological activity, and the depth of water.

The continental shelf is the gently sloping area directly adjacent to the continent and consists of both a high energy and a low energy environment. The high energy inner shelf is located closest to the shoreline; in this region sediment is constantly stirred up by wave action and tidal currents which produce large underwater cross-bedded dunes called a hummocky terrain. Further from shore, in the lower energy outer shelf environment finer sediments, such as silts and clays, are deposited.

As sediment moves across the continental shelf it eventually comes to rest at the continental slope and rise. Within this region sediment is transported down the continental slope in turbidity currents a turbidity current is simply sediment laden current of water; on land this would be most similar to a mudflow. As a result of the constant deposition by turbidity currents graded bedding is common.

Beyond the continental shelf the sea floor is covered by pelagic clays. The energy of this environment is too low to transport larger particles such as sand and gravels from the continent. Instead the source of most detrital sediment is dust blown by wind off the continents or oceanic islands, this sediment is known as pelagic clay. Along with pelagic clay the calcareous ($CaCO_3$) and Siliceous ($SiO_2$) skeletal remains of microscopic organisms are also deposited in the form of oozes.

Limestone and dolostone are the most common carbonate rocks and we now know that dolostone is an altered form of limestone. In some respect the deposition of limestone is very similar to the deposition of detrital (clastic) sedimentary rock as it is often formed through the continuous deposition of sand-sized grains and microcrystalline carbonate muds. Because of this they may also contain sedimentary structures such as crossbedding and ripple marks. However, unlike clastic sedimentary rocks, carbonates are produced directly within their environment through chemical precipitation. During this process positively-charged calcium ions bond with negatively-charged carbonate ions to produce solid crystals of calcium carbonate (calcite) that precipitate out on the sea floor. This process requires relatively warm shallow water as calcium carbonate dissolves in cold-deep water.

# References

- Sedimentology, entry: newworldencyclopedia.org, Retrieved 20 June, 2019

- Gehling, James; Jensen, Sören; Droser, Mary; Myrow, Paul; Narbonne, Guy (March 2001). "Burrowing below the basal Cambrian GSSP, Fortune Head, Newfoundland". Geological Magazine. 138 (2): 213–218. Doi:10.1017/S001675680100509X

- Sedimentary-facies, science: britannica.com, Retrieved 21 July, 2019

- Schlager, Wolfgang (2005). Carbonate sedimentology and sequence stratigraphy. SEPM Concepts in Sedimentology and Paleontology. ISBN 978-1565761162

- Stratigraphy-geology, science: britannica.com, Retrieved 22 August, 2019

- Smith, J. B.; Lamanna, M.C.; Lacovara, K.J.; Dodson, P. Jr.; Poole, J.C.; Giegengack, R. (1 June 2001). "A Giant Sauropod Dinosaur from an Upper Cretaceous Mangrove Deposit in Egypt". Science. 292 (5522): 1704–1706. Doi:10.1126/science.1060561. ISSN 1095-9203. PMID 11387472

- Description-of-lithostratigraphy: geologylearn.blogspot.com, Retrieved 23 January, 2019

# 3

# Sedimentation

The tendency of particles in suspension to settle out of the fluid in which they are entrained and come to rest against a barrier is known as sedimentation. Some of the processes which are studied in relation to the process of sedimentation are fluvial processes and aeolian processes. This chapter has been carefully written to provide an easy understanding sedimentation and related processes.

Sedimentation, in the geological sciences, is a process of deposition of a solid material from a state of suspension or solution in a fluid (usually air or water). Broadly defined it also includes deposits from glacial ice and those materials collected under the impetus of gravity alone, as in talus deposits, or accumulations of rock debris at the base of cliffs. The term is commonly used as a synonym for sedimentary petrology and sedimentology.

The physics of the most common sedimentation process, the settling of solid particles from fluids, has long been known. The settling velocity equation formulated in 1851 by G.G. Stokes is the classic starting point for any discussion of the sedimentation process. Stokes showed that the terminal settling velocity of spheres in a fluid was inversely proportional to the fluid's viscosity and directly proportional to the density difference of fluid and solid, the radius of the spheres involved, and the force of gravity. Stokes' equation is valid, however, only for very small spheres (under 0.04 millimetre [0.0015 inch] in diameter) and hence various modifications of Stokes' law have been proposed for nonspherical particles and particles of larger size.

No settling velocity equation, however valid, provides a sufficient explanation of even the basic physical properties of natural sediments. The grain size of the clastic elements and their sorting, shape, roundness, fabric, and packing are the results of complex processes related not only to the density and viscosity of the fluid medium but also to the translational velocity of the depositing fluid, the turbulence resulting from this motion, and the roughness of the beds over which it moves. These processes also are related to various mechanical properties of the solid materials propelled, to the duration of sediment transport, and to other little-understood factors.

Sedimentation is generally considered by geologists in terms of the textures, structures, and fossil content of the deposits laid down in different geographic and geomorphic environments. Great efforts have been made to differentiate between continental, near-shore, marine, and other deposits in the geologic record. The classification of environments and criteria for their recognition is still a subject of lively debate. The analysis and interpretation of ancient deposits has been advanced by the study of modern sedimentation. Oceanographic and limnologic expeditions have shed much light on sedimentation in the Gulf of Mexico, the Black Sea, and the Baltic Sea, and in various estuaries, lakes, and fluvial basins in all parts of the world.

Chemical sedimentation is understood in terms of chemical principles and laws. Although the famous physical chemist J. H. van't Hoff applied the principles of phase equilibria to the problem of crystallizing brines and the origin of salt deposits as early as 1905, little effort was made to apply physical chemistry to the problems of chemical sedimentation. More recently, however, there has been investigation of the role of the redox (mutual reduction and oxidation) potential and pH (acidity–alkalinity) in the precipitation of many chemical sediments, and a renewed effort has been made to apply known thermodynamic principles to the origin of anhydrite and gypsum deposits, to the chemistry of dolomite formation, and to the problem of the ironstones and related sediments.

The geochemist also considers the sedimentation process in terms of the chemical end products. To him sedimentation is like a gigantic chemical analysis in which the primary constituents of the Earth's silicate crust are separated from one another in a manner similar to that achieved in the course of a quantitative analysis of rock material in the laboratory. The results of this chemical fractionation are not always perfect, but by and large the results are remarkably good. Geochemical fractionation, which began in Precambrian time, has resulted in an enormous accumulation of sodium in the sea, calcium and magnesium in the limestones and dolomites, silicon in the bedded cherts and orthoquartzitic sandstones, carbon in the carbonates and carbonaceous deposits, sulfur in the bedded sulfates, iron in the ironstones, and so on. Although magmatic segregation has, in some instances, produced monomineralic rocks such as dunite and pyroxenite, no igneous or metamorphic process can match the sedimentation process in effective isolation and concentration of these and other elements.

The main factors that control the breakdown of rocks into sediments are:

- Climate,

- Topography,

- Vegetation,

- Properties (physical and chemical) of the rock.

The particles that are broken down are called sediments. Sediments are classified according to their size, ranging from silts and muds up to gravels and boulders.

Sediments can then be transported from their source, often to great distances. The main factors that control the transportation of sediments are:

- Water,

- Wind (particularly in arid regions),

- Gravity (with all sediments flowing downhill regardless of the slope).

## Principle of Uniformitarianism

The principle of uniformitarianism is that processes which operate on the Earth's surface today are similar to those that operated in the past. It is a fundamental principle in sedimentary geology and was first proposed by Charles Lyell in 1830. Using this principle when studying present-day sedimentary environments (e.g. coral reef systems, delta systems, river systems), we can determine fundamental principles such as:

- Rates of sedimentation,

- Geometry of sediment sequences,

- Rates of compaction,

- Amount of water present in the sediments,

which can then be applied to much older sedimentary rock sequences.

## Stratification

Horizontal layering in sedimentary rocks is called bedding or stratification. It forms by the settling of particles from either water or air . Layer boundaries are natural planes of weakness along which the rocks can break and fluids can flow. As long as the sequence of layers has not been deformed or overturned, the youngest layers are at the top and the oldest are at the bottom. This sequence of stratification is the basis for the stratigraphic time scale. These observations were first made by a Danish physician, Nicolaus Steno, who in 1669 formulated the principles of horizontality, superposition (younger layers on top of older ones) and original continuity (sedimentary layers represent former continuous sheets).

## Preservation of Sedimentary Sequences

Most sedimentary sequences that are preserved in the rock record are formed from catastrophic deposition such as floods, mud flows, rock slides and melting of glaciers. For a sediment sequence to be preserved and lithified (turned into rock), it must be covered over by younger sediments soon after it is deposited and water within the sequence must be expelled (this usually achieved through compaction by the weight of overlying layers).

## Weathering and Erosion

Weathering is the process where rocks break down under the effects of water and air. It consists of two processes which always act together:

- Fragmentation (known as mechanical or physical weathering).

- Decay (known as chemical weathering).

Erosion is the process of the movement of weathering products, by water and air.

The smaller the pieces (or fragments), the greater the surface area available for chemical attack and the faster the pieces decay. Important agents in weathering are rainfall, wind, ice, snow, rivers, seawater, vegetation and living organisms.

Soil is both a factor in weathering and the result of it. Once soil starts to form, rock weathers more rapidly and more soil is formed. Rainwater, which is mainly $H_2O$, also contains small amounts of dissolved $CO_2$ and $H_2S$ which react strongly with many rock materials.

## Weathering and Feldspars

Feldspar is the most common mineral in many igneous and metamorphic rocks, particularly in granites. In temperate climates, granites are quite resistant to erosion whereas in humid to tropical regions, granites decay easily. This is because in areas of greater rainfall, the feldspars decay into clay minerals that are then easily released from the rock. Feldspars generally only remain untouched in very arid climates.

The most common clay mineral formed is kaolinite. In the weathering of feldspars to kaolinite, potassium, sodium, calcium and silicon are released from the feldspar structure and go into solution. Since the speed of chemical reactions increases with increasing temperature, weathering occurs at a much faster rate in tropical areas than in more temperate ones. Under extreme weathering conditions in tropical climates, the clays themselves decay further into the mineral gibbsite $(Al\,(OH)_3)$ the main ore of aluminium. The resulting rock is a bauxite or laterite.

## Weathering and Carbonates

Whereas feldspar minerals decay into clays, carbonate minerals can completely dissolve. Caves are a characteristic product of the weathering of limestone (rocks made up predominantly of the carbonate minerals calcite and dolomite) in humid climates.

The main agent of weathering in this case is groundwater containing dissolved $CO_2$ (carbon dioxide). This dissolution of limestone results in the enrichment of the groundwater in Ca (calcium) ions, leading to what is known as 'hard water'. $CO_2$ is extracted from the atmosphere and from organic material in the soil and dissolves in the groundwaters. The greater the weathering of limestone, the more $CO_2$ is removed from the atmosphere. As limestone dissolves faster than silicate rocks, the chemical weathering of limestone accounts for more of the total chemical erosion of the land surface than any other rocks, even though much larger areas of the Earth consist of silicate rocks.

## Resistance of Minerals to Weathering

Measurements of weathering in the field can be combined with experiments in the laboratory to determine the relative resistance of minerals to weathering. The products of weathering are generally more resistant to further weathering than other minerals, particularly iron oxide and clay minerals. As quartz is generally insoluble and chemically stable, it tends not to weather readily under most conditions. The order of mineral stability under weathering conditions is related to the stabilities of chemical bonds and crystal structures under different temperatures and pressures.

## Weathering in different Rock types

Layered rocks break into slabs or plates along bedding planes whereas massive rocks break along regularly spaced planar cracks, called joints. In some igneous rocks, the joints take the form of sheets - sets of parallel closely-spaced planar surfaces. In areas of large daily temperature gradients, thermal expansion often accompanies frost action and chemical weathering.

Exfoliation is the peeling-off of large curved sheets or slabs of rock from the weathering surface of an outcrop. Spheroidal weathering is a similar phenomenon in which rounded boulders split off layers or shells from the weathered surface.

## Shapes of Weathered Fragments

The shapes of fragments caused by weathering and erosion are largely inherited from the patterns of joints, bedding and other structures in the parent rock rather than being produced by transport. The size of fragments is a good clue to the intensity of mechanical erosion. In general, the higher or steeper the landscape, the larger the fragments. Once fragments have been moved from their source rock, they enter new environments of weathering and erosion and undergo further breakdown. Once boulders fall into streams they break and abrade quickly, and the size of pebbles downstream decreases rapidly with distance from the source. The different sizes and shapes of eroded particles (from huge boulders to clay-sized particles) can be attributed to the characteristics of their source rocks and their distance from the source.

## Rates of Erosion

Rates of erosion can be averaged over regions and are called rates of denudation(measured in mm per thousand years). These rates are greatest in valley glaciers and badlands (deeply eroded areas) and lowest in areas of low relief and temperate and rainforest regions. The rate of denudation is primarily controlled by topography and climate. Human influences accelerate the rate of denudation by three to ten times, with the highest figures being recorded in areas of intensive land use.

One of the highest rates of denudation measured is in the Tamur Basin of the Himalayas. The combination of steep slopes, unconsolidated material (sediment), a glaciated terrain and human modification has resulted in a rate of 4700 mm per 1000 years.

## Organic Matter and Sedimentary Processes

The biosphere (all biological activity such as plants, animals, and their remains) also plays a vital role in sedimentary processes. All organic matter eventually decomposes, releasing vital nutrients (such as N, Ca, C) into the soil and sea. Both coal and oil are formed by the interaction of buried organic matter with sedimentary processes.

## Diagenesis

Diagenesis is the alteration of the mineralogy and texture of sediments at low temperatures and pressures. It affects sediments close to the Earth's surface. There are two main processes operating:

- Compaction: by overlying sediments, involving the close-packing of the individual grains by eliminating the pore space and expulsion of entrapped water.

- Cementation: development of secondary material in the former pore spaces which then binds the sedimentary particles together. This material may be introduced from the passage of groundwater or derived from solution.

# GRAIN SIZE AND COMPOSITION

The grain sizes of sediments and sedimentary rocks are a matter of great interest to geologists.

Different size sediment grains form different types of rocks and can reveal information about the landform and environment of an area from millions of years prior.

## Types of Sediment Grains

Sediments are classified by their method of erosion as either clastic or chemical. Chemical sediment is broken down through chemical weathering with transportation, a process known as corrosion, or without. That chemical sediment is then suspended in a solution until it precipitates. Think of what happens to a glass of saltwater that has been sitting out in the sun.

Clastic sediments are broken down through mechanical means, like abrasion from wind, water or ice. They are what most people think of when mentioning sediment; things like sand, silt, and clay. Several physical properties are used to describe sediment, like shape (sphericity), roundness and grain size.

Of these properties, grain size is arguably the most important. It can help a geologist interpret the geomorphic setting (both present and historical) of a site, as well as whether the sediment was transported there from regional or local settings. Grain size determines just how far a piece of sediment can travel before coming to a halt.

Clastic sediments form a wide range of rocks, from mudstone to conglomerate, and soil depending on their grain size. Within many of these rocks, the sediments are clearly distinguishable--especially with a little help from a magnifier.

## Sediment Grain Sizes

The Wentworth scale was published in 1922 by Chester K. Wentworth, modifying an earlier scale by Johan A. Udden. Wentworth's grades and sizes were later supplemented by William Krumbein's phi or logarithmic scale, which transforms the millimeter number by taking the negative of its logarithm in base 2 to yield simple whole numbers. The following is a simplified version of the much more detailed USGS version.

| Millimeters | Wentworth Grade | Phi ($\Phi$) Scale |
|---|---|---|
| >256 | Boulder | −8 |
| >64 | Cobble | −6 |
| >4 | Pebble | −2 |
| >2 | Granule | −1 |
| >1 | Very coarse sand | 0 |
| >1/2 | Coarse sand | 1 |
| >1/4 | Medium sand | 2 |
| >1/8 | Fine sand | 3 |
| >1/16 | Very fine sand | 4 |
| >1/32 | Coarse silt | 5 |
| >1/64 | Medium silt | 6 |
| >1/128 | Fine silt | 7 |
| >1/256 | Very fine silt | 8 |
| <1/256 | Clay | >8 |

The size fraction larger than sand (granules, pebbles, cobbles. and boulders) is collectively called gravel, and the size fraction smaller than sand (silt and clay) is collectively called mud.

## Clastic Sedimentary Rocks

Sedimentary rocks form whenever these sediments are deposited and lithified and can be classified based on the size of their grains.

- Gravel forms coarse rocks with grains over 2 mm in size. If the fragments are rounded, they form conglomerate, and if they are angular, they form breccia.

- Sand, as you may guess, forms sandstone. Sandstone is medium-grained, meaning its fragments are between 1/16 mm and 2 mm.

- Silt forms fine-grained siltstone, with fragments between 1/16 mm and 1/256 mm.

- Anything less than 1/256 mm results in either claystone or mudstone. Two types of mudstone are shale and argillite, which is shale that has undergone very low-grade metamorphism.

Geologists determine grain sizes in the field using printed cards called comparators, which usually have a millimeter scale, phi scale, and angularity chart. They are especially useful for larger sediment grains. In the laboratory, comparators are supplemented by standard sieves.

# FLUVIAL PROCESSES

Fluvial process is the physical interaction of flowing water and the natural channels of rivers and streams. Such processes play an essential and conspicuous role in the denudation of land surfaces and the transport of rock detritus from higher to lower levels.

Over much of the world the erosion of landscape, including the reduction of mountains and the building of plains, is brought about by the flow of water. As the rain falls and collects in watercourses, the process of erosion not only degrades the land, but the products of erosion themselves become the tools with which the rivers carve the valleys in which they flow. Sediment materials eroded from one location are transported and deposited in another, only to be eroded and redeposited time and again before reaching the ocean. At successive locations, the riverine plain and the river channel itself are products of the interaction of a water channel's flow with the sediments brought down from the drainage basin above.

The velocity of a river's flow depends mainly upon the slope and the roughness of its channel. A steeper slope causes higher flow velocity, but a rougher channel decreases it. The slope of a river corresponds approximately to the fall of the country it traverses. Near the source, frequently in hilly regions, the slope is usually steep, but it gradually flattens out, with occasional irregularities, until, in traversing plains along the latter part of the river's course, it usually becomes quite mild. Accordingly, large streams usually begin as torrents with highly turbulent flow and end as gently flowing rivers.

In floodtime, rivers bring down large quantities of sediment, derived mainly from the disintegration of the surface layers of the hills and valley slopes by rain and from the erosion of the riverbed by flowing water. Glaciers, frost, and wind also contribute to the disintegration of the Earth's surface and to the supply of sediment to rivers. The power of a river current to transport materials depends to a large extent on its velocity, so that torrents with a rapid fall near the sources of rivers can carry down rocks, boulders, and large stones. These are gradually ground by attrition in their onward course into shingle, gravel, sand, and silt and are carried forward by the main river toward the sea or partially strewn over flat plains during floods. The size of the materials deposited in the bed of the river becomes smaller as the reduction of velocity diminishes the transporting power of the current.

Since the earliest days of modern applied hydraulics, engineering research has attempted to better understand sediment transportation. Because sediment particles are generally heavier than the amount of water they displace, the Archimedes principle could not be used to explain the fact that heavy sediment was capable of being lifted and transported by flowing water. Another explanation was, consequently, required. Twentieth-century research distinguishes, in this connection, between "bed load" on the one hand and "suspended load" on the other. The former is composed of the larger particles, which are either rolled or pushed along the bed of the stream or which "jump," or saltate, from the crest of one ripple to another if the velocity is sufficiently great. On the other hand, the smaller particles, the suspended sediment once picked up and lifted by the moving water, may remain in suspension for considerable periods of time and thus be transported over many kilometres.

Exner Equation:

The Exner equation is a statement of conservation of mass that applies to sediment in a fluvial system such as a river. It was developed by the Austrian meteorologist and sedimentologist Felix Maria Exner, from whom it derives its name.

Equation:

The Exner equation describes conservation of mass between sediment in the bed of a channel and sediment that is being transported. It states that bed elevation increases (the bed aggrades) proportionally to the amount of sediment that drops out of transport, and conversely decreases (the bed degrades) proportionally to the amount of sediment that becomes entrained by the flow.

Basic Equation:

The equation states that the change in bed elevation, $\eta$, over time, $t$, is equal to one over the grain packing density, $\varepsilon_o$, times the negative divergence of sediment flux, $\mathbf{q_s}$.

$$\frac{\partial \eta}{\partial t} = -\frac{1}{\varepsilon_o} \nabla \cdot \mathbf{q_s}$$

Note that $\varepsilon_o$ can also be expressed as $(1-\lambda_p)$, where $\lambda_p$ equals the bed porosity.

Good values of $\varepsilon_o$ for natural systems range from 0.45 to 0.75. A typical good value for spherical grains is 0.64, as given by random close packing. An upper bound for close-packed spherical grains is 0.74048. This degree of packing is extremely improbable in natural systems, making random close packing the more realistic upper bound on grain packing density.

Often, for reasons of computational convenience and/or lack of data, the Exner equation is used in its one-dimensional form. This is generally done with respect to the down-stream direction $x$, as one is typically interested in the down-stream distribution of erosion and deposition though a river reach.

$$\frac{\partial \eta}{\partial t} = -\frac{1}{\varepsilon_o}\frac{\partial \mathbf{q_s}}{\partial x}$$

### Including External Changes in Elevation

An additional form of the Exner equation adds a subsidence term, $\sigma$, to the mass-balance. This allows the absolute elevation of the bed $\eta$ to be tracked over time in a situation in which it is being changed by outside influences, such as tectonic or compression-related subsidence (isostatic compression or rebound). In the convention of the following equation, $\sigma$ is positive with an increase in elevation over time and is negative with a decrease in elevation over time.

$$\frac{\partial \eta}{\partial t} = -\frac{1}{\varepsilon_o}\nabla \cdot \mathbf{q_s} + \sigma$$

# AEOLIAN PROCESSES

Wind erosion of soil at the foot of Chimborazo, Ecuador.

Rock carved by drifting sand below Fortification Rock in Arizona.

Aeolian processes, also spelled eolian or æolian, pertain to wind activity in the study of geology and weather and specifically to the wind's ability to shape the surface of the Earth (or other planets). Winds may erode, transport, and deposit materials and are effective agents in regions with sparse vegetation, a lack of soil moisture and a large supply of unconsolidated sediments. Although water is a much more powerful eroding force than wind, aeolian processes are important in arid environments such as deserts.

The term is derived from the name of the Greek god Aeolus, the keeper of the winds.

## Wind Erosion

A rock sculpted by wind erosion in the Altiplano region of Bolivia.

Sand blowing off a crest in the Kelso Dunes of the Mojave Desert, California.

Wind-carved alcove in the Navajo Sandstone near Moab, Utah.

Wind erodes the Earth's surface by deflation (the removal of loose, fine-grained particles by the turbulent action of the wind) and by abrasion (the wearing down of surfaces by the grinding action and sandblasting by windborne particles).

Regions which experience intense and sustained erosion are called deflation zones. Most aeolian deflation zones are composed of desert pavement, a sheet-like surface of rock fragments that remains after wind and water have removed the fine particles. Almost half of Earth's desert surfaces are stony deflation zones. The rock mantle in desert pavements protects the underlying material from deflation.

A dark, shiny stain, called desert varnish or rock varnish, is often found on the surfaces of some desert rocks that have been exposed at the surface for a long period of time. Manganese, iron oxides, hydroxides, and clay minerals form most varnishes and provide the shine.

Deflation basins, called blowouts, are hollows formed by the removal of particles by wind. Blowouts are generally small, but may be up to several kilometers in diameter.

Wind-driven grains abrade landforms. In parts of Antarctica wind-blown snowflakes that are technically sediments have also caused abrasion of exposed rocks. Grinding by particles carried in the wind creates grooves or small depressions. Ventifacts are rocks which have been cut, and sometimes polished, by the abrasive action of wind.

Sculpted landforms, called yardangs, are up to tens of meters high and kilometers long and are forms that have been streamlined by desert winds. The famous Great Sphinx of Giza in Egypt may be a modified yardang.

## List of Major Aeolian Movements

Major global aeolian dust movements thought to influence and/or be influenced by weather and climate variation:

- From Sahara (specifically Sahel and Bodélé Depression) averaged 182 million tons of dust each year between 2007 and 2011 and carry it past the western edge of the Sahara at longitude 15W. Variation: 86% (2007/11). Destination: 132 mln tons cross the Atlantic (ave), 27.7 mln tons fall in Amazon Basin (ave), 43 mln make it to the Caribbean. Texas and Florida also receive the dust. Events have become far more common in recent decades. Source: NASA's Cloud-Aerosol Lidar and Infrared Pathfinder Satellite Observation (CALIPSO) data. Harmattan winter dust storms in West Africa also occur blowing dust to the ocean.

- Gobi Desert to Korea, Japan, Taiwan (at times) and even Western USA (blowing east).

- Thar Desert pre-monsoon towards Delhi, Uttar Pradesh, Indo-Gangetic Plain.

- Shamal June–July winds blowing dust in primarily north to south in Saudi Arabia, Iran, Iraq, UAE, and parts of Pakistan.

- Haboob dust storms in Sudan, Australia, Arizona associated with monsoon.

- Khamsin dust from Libya, Egypt and Levant in Spring associated with extratropical cyclones.

- Dust Bowl event in USA, carried sand eastward. 5500 tons were deposited in Chicago area.

- Sirocco sandy winds from Africa/Sahara blowing north into South Europe.

- Kalahari Desert blowing east to southern Indian Ocean and Australia.

## Transport

Dust storm approaching Spearman, Texas.

A massive sand storm cloud is about to envelop a military camp as it rolls over Al Asad, Iraq.

Particles are transported by winds through suspension, saltation (skipping or bouncing) and creeping (rolling or sliding) along the ground.

Small particles may be held in the atmosphere in suspension. Upward currents of air support the weight of suspended particles and hold them indefinitely in the surrounding air. Typical winds near Earth's surface suspend particles less than 0.2 millimeters in diameter and scatter them aloft as dust or haze.

Saltation is downwind movement of particles in a series of jumps or skips. Saltation normally lifts sand-size particles no more than one centimeter above the ground and proceeds at one-half to one-third the speed of the wind. A saltating grain may hit other grains that jump up to continue the saltation. The grain may also hit larger grains that are too heavy to hop, but that slowly creep forward as they are pushed by saltating grains. Surface creep accounts for as much as 25 percent of grain movement in a desert.

Aeolian turbidity currents are better known as dust storms. Air over deserts is cooled significantly when rain passes through it. This cooler and denser air sinks toward the desert surface. When it reaches the ground, the air is deflected forward and sweeps up surface debris in its turbulence as a dust storm.

Crops, people, villages, and possibly even climates are affected by dust storms. Some dust storms are intercontinental, a few may circle the globe, and occasionally they may engulf entire planets. When the Mariner 9 spacecraft entered its orbit around Mars in 1971, a dust storm lasting one month covered the entire planet, thus delaying the task of photo-mapping the planet's surface.

Most of the dust carried by dust storms is in the form of silt-size particles. Deposits of this wind-blown silt are known as loess. The thickest known deposit of loess, 335 meters, is on the Loess Plateau in China. This very same Asian dust is blown for thousands of miles, forming deep beds in places as far away as Hawaii. In Europe and in the Americas, accumulations of loess are generally from 20 to 30 meters thick. The soils developed on loess are generally highly productive for agriculture.

Aeolian transport from deserts plays an important role in ecosystems globally, e.g. by transport of minerals from the Sahara to the Amazon basin. Saharan dust is also responsible for forming red clay soils in southern Europe. Aeolian processes are affected by human activity, such as the use of 4x4 vehicles.

Small whirlwinds, called dust devils, are common in arid lands and are thought to be related to very intense local heating of the air that results in instabilities of the air mass. Dust devils may be as much as one kilometer high.

## Deposition

Wind-deposited materials hold clues to past as well as to present wind directions and intensities. These features help us understand the present climate and the forces that molded it. Wind-deposited sand bodies occur as sand sheets, ripples, and dunes.

Sand sheets are flat, gently undulating sandy plots of sand surfaced by grains that may be too large for saltation. They form approximately 40 percent of aeolian depositional surfaces. The Selima Sand Sheet in the eastern Sahara Desert, which occupies 60,000 square kilometers in southern Egypt and northern Sudan, is one of the Earth's largest sand sheets. The Selima is absolutely flat in a few places; in others, active dunes move over its surface.

Wind blowing on a sand surface ripples the surface into crests and troughs whose long axes are perpendicular to the wind direction. The average length of jumps during saltation corresponds to the wavelength, or distance between adjacent crests, of the ripples. In ripples, the coarsest materials collect at the crests causing inverse grading. This distinguishes small ripples from dunes, where the coarsest materials are generally in the troughs. This is also a distinguishing feature between water laid ripples and aeolian ripples.

Accumulations of sediment blown by the wind into a mound or ridge, dunes have gentle upwind slopes on the windward side. The downwind portion of the dune, the lee slope, is commonly a steep avalanche slope referred to as a slipface. Dunes may have more than one slipface. The minimum height of a slipface is about 30 centimeters.

Wind-blown sand moves up the gentle upwind side of the dune by saltation or creep. Sand accumulates at the brink, the top of the slipface. When the buildup of sand at the brink exceeds the angle of repose, a small avalanche of grains slides down the slipface. Grain by grain, the dune moves downwind.

Some of the most significant experimental measurements on aeolian sand movement were performed by Ralph Alger Bagnold, a British army engineer who worked in Egypt prior to World War II. Bagnold investigated the physics of particles moving through the atmosphere and deposited by wind. He recognized two basic dune types, the crescentic dune, which he called "barchan", and the linear dune, which he called longitudinal or "seif" (Arabic for "sword").

A 2011 study published in *Catena* examined the effect of vegetation on aeolian dust accumulation in the semiarid steppe of northern China. Using a series of trays with different vegetation coverage and a control model with none, the authors found that an increase in vegetation coverage improves the efficiency of dust accumulation and adds more nutrients to the environment, particularly organic carbon. Two critical point were revealed by their data: the efficiency of trapping dust increases slowly above 15% coverage, and decreases rapidly below 15% coverage. at around 55%-75% coverage, dust accumulation reaches a maximum capacity.

In Europe, the European Commission requested the Joint Research Centre to develop the first pan-European wind erosion map. In a first step, a group of scientists have used the LUCAS topsoil dataset to develop the wind erosion susceptibility of European soils. Then, they have developed an index to land susceptibility for making a qualitative assessment of wind erosion. Finally, they modified the RWEQ model to estimate the soil Loss Due to Wind Erosion in European Agricultural Soils.

A three-year quantitative study on the effects of vegetation removal on wind erosion found that the removal of grasses in an aeolian environment increased the rate of soil deposition. In the same study, a relationship was shown between decreasing plant density with decreasing soil nutrients. Similarly, horizontal soil flux across the test site was shown to increase with increasing vegetation removal.

A 1998 study published in Earth Surfaces Processes and Landforms investigated the relationship between vegetative cover on sand surfaces with the rate of sand transport. It was found that sand flux decreased exponentially with vegetation cover. This was done by measuring plots of land with varying degrees of vegetation against rates of sand transport. The authors contend that this relationship can be utilized to manipulate rates of sediment flux by introducing vegetation in an area or to quantify human impact by recognizing vegetation loss's effect on sandy landscapes.

Cross-bedding of sandstone near Mount Carmel road, Zion National Park, indicating wind action and sand dune formation prior to formation of rock.

Mesquite Flat Dunes in Death Valley looking toward the Cottonwood Mountains from the north west arm of Star Dune.

Holocene eolianite deposit on Long Island, The Bahamas. This unit is formed of wind-blown carbonate grains.

# GLACIAL DEPOSITION

Sediments transported and deposited during the Pleistocene glaciations are important sources of construction materials and are valuable as reservoirs for groundwater. Because they are almost all unconsolidated, they have significant implications for mass wasting.

Figure illustrates some of the ways that sediments are transported and deposited. The Bering Glacier is the largest in North America, and although most of it is in Alaska, it flows from an icefield that extends into southwestern Yukon. The surface of the ice is partially, or in some cases completely, covered with rocky debris that has fallen from surrounding steep rock faces. There are muddy rivers issuing from the glacier in several locations, depositing sediment on land, into Vitus Lake, and directly into the ocean. There are dirty icebergs shedding their sediment into the lake. And, not visible in this view, there are sediments being moved along beneath the ice.

Part of the Bering Glacier in southeast Alaska, the largest glacier in North America. It is about 14 km across in the centre of this view.

The formation and movement of sediments in glacial environments is shown diagrammatically in figure. There are many types of glacial sediment generally classified by whether they are transported on, within, or beneath the glacial ice. The main types of sediment in a glacial environment are described below.

Supraglacial (on top of the ice) and englacial (within the ice) sediments that slide off the melting front of a stationary glacier can form a ridge of unsorted sediments called an end moraine. The end moraine that represents the farthest advance of the glacier is a terminal moraine. Sediments transported and deposited by glacial ice are known as till.

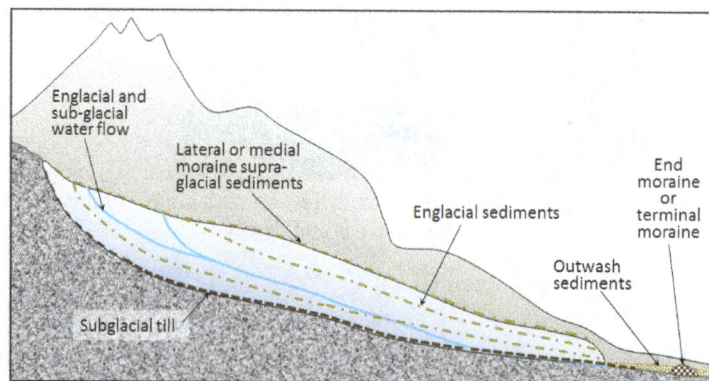

A depiction of the various types of sediments associated with glaciation. The glacier is shown in cross-section.

Subglacial sediment (e.g., lodgement till) is material that has been eroded from the underlying rock by the ice, and is moved by the ice. It has a wide range of grain sizes, including a relatively high proportion of silt and clay. The larger clasts (pebbles to boulders in size) tend to become partly rounded by abrasion. When a glacier eventually melts, the lodgement till is exposed as a sheet of well-compacted sediment ranging from several centimetres to many metres in thickness. Lodgement till is normally unbedded. An example is shown in figure a.

Supraglacial sediments are primarily derived from freeze-thaw eroded material that has fallen onto the ice from rocky slopes above. These sediments form lateral moraines and, where two glaciers meet, medial moraines. (Medial moraines are visible on the Aletsch Glacier in figure.) Most of this material is deposited on the ground when the ice melts, and is therefore called ablation till, a mixture of fine and coarse angular rock fragments, with much less sand, silt, and clay than lodgement till. An example is shown in figure b. When supraglacial sediments become incorporated into the body of the glacier, they are known as englacial sediments.

Examples of glacial till: a: lodgement till from the front of the Athabasca Glacier, Alberta; b: ablation till at the Horstman Glacier, Blackcomb Mountain, B.C. [SE]

Massive amounts of water flow on the surface, within, and at the base of a glacier, even in cold areas and even when the glacier is advancing. Depending on its velocity, this water is able to move sediments of various sizes and most of that material is washed out of the lower end of the glacier and deposited as outwash sediments. These sediments accumulate in a wide range of environments in the proglacial region (the area in front of a glacier), most in fluvial environments, but some in lakes and the ocean. Glaciofluvial sediments are similar to sediments deposited in normal fluvial environments, and are dominated by silt, sand, and gravel. The grains tend to be moderately well rounded, and the sediments have similar sedimentary structures (e.g., bedding, cross-bedding, clast imbrication) to those formed by non-glacial streams.

Examples of glaciofluvial sediments: a: glaciofluvial sand of the Quadra Sand Formation at Comox, B.C.; b: glaciofluvial gravel and sand, Nanaimo, B.C.

A large proglacial plain of sediment is called a sandur (a.k.a. an outwash plain), and within that area, glaciofluvial deposits can be tens of metres thick. In situations where a glacier is receding, a block of ice might become separated from the main ice sheet and become buried in glaciofluvial sediments. When the ice block eventually melts, a depression forms, known as a kettle, and if this fills with water, it is known as a kettle lake .

A kettle lake amid vineyards and orchards in the Osoyoos area of B.C.

A subglacial stream will create its own channel within the ice, and sediments that are being transported and deposited by the stream will build up within that channel. When the ice recedes, the sediment will remain to form a long sinuous ridge known as an esker. Eskers are most common in areas of continental glaciation. They can be several metres high, tens of metres wide, and tens of kilometres long .

Part of an esker that formed beneath the Laurentide Ice Sheet in northern Canada.

Outwash streams commonly flow into proglacial lakes where glaciolacustrine sediments are deposited. These are dominated by silt- and clay-sized particles and are typically laminated on the millimetre scale. In some cases, varves develop; varves are series of beds with distinctive summer and winter layers: relatively coarse in the summer when melt discharge is high, and finer in the winter, when discharge is very low. Icebergs are common on proglacial lakes, and most of them contain englacial sediments of various sizes. As the bergs melt, the released clasts sink to the bottom and are incorporated into the glaciolacustrine layers as drop stones.

The processes that occur in proglacial lakes can also take place where a glacier terminates in the ocean. The sediments deposited there are called glaciomarine sediments.

Examples of glacial sediments formed in quiet water: a: glaciolacustrine sediment with a drop stone, Nanaimo, B.C.; and b: a laminated glaciomarine sediment, Englishman River, B.C. Although not visible in this photo, the glaciomarine sediment has marine shell fossils.

# References

- Sedimentation-geology, science: britannica.com, Retrieved 24 February, 2019

- Paola, C.; Voller, V. R. (2005). "A generalized Exner equation for sediment mass balance". Journal of Geophysical Research. 110: F04014. Bibcode:2005JGRF..11004014P. Doi:10.1029/2004JF000274

- Sedimentary-processes, shaping-earth, minerals, learn: australianmuseum.net.au, Retrieved 25 March, 2019

- Orgiazzi, A.; Ballabio, C.; Panagos, P.; Jones, A.; Fernández-Ugalde, O. (2018). "LUCAS Soil, the largest expandable soil dataset for Europe: a review". European Journal of Soil Science. 69: 140–153. Doi:10.1111/ejss.12499. ISSN 1365-2389

- All-about-sediment-grain-size-1441194: thoughtco.com, Retrieved 26 April, 2019

- Orgiazzi, A.; Ballabio, C.; Panagos, P.; Jones, A.; Fernández-Ugalde, O. (2018). "LUCAS Soil, the largest expandable soil dataset for Europe: a review". European Journal of Soil Science. 69: 140–153. doi:10.1111/ejss.12499. ISSN 1365-2389

- Fluvial-process, science: britannica.com, Retrieved 27 May, 2019

- Borrelli, Pasquale; Panagos, Panos; Ballabio, Cristiano; Lugato, Emanuale; Weynants, Melanie; Montanarella, Luca (2016-05-01). "Towards a Pan-European Assessment of Land Susceptibility to Wind Erosion". Land Degradation & Development. 27 (4): 1093–1105. Doi:10.1002/ldr.2318. ISSN 1099-145X

- 16-4-glacial-deposition, chapter, geology: opentextbc.ca, Retrieved 28 June, 2019

# 4

# Sedimentary Structures and Sedimentary Rocks

Rocks that are formed by the deposition or accumulation of small particles are known as sedimentary rocks. The large three dimensional physical features of sedimentary rocks are known as sedimentary structures. This chapter discusses in detail various types of sedimentary structures and rocks as well as their features.

## SEDIMENTARY STRUCTURES

Sedimentary structures are visible features within sedimentary rocks that formed at the time of deposition and represent manifestations of the physical and biological processes that operated in depositional environments.

Through careful observation over the past few centuries, geologists have discovered that the accumulation of sediments and sedimentary rocks takes place according to some important geological principles, as follows:

- The principle of original horizontality states that sediments accumulate in essentially horizontal layers. The implication is that tilted sedimentary layers observed to day must have been subjected to tectonic forces.

- The principle of superposition states that sedimentary layers are deposited in sequence, and that unless the entire sequence has been turned over by tectonic processes, the layers at the bottom are older than those at the top.

- The principle of inclusions states that any rock fragments in a sedimentary layer must be older than the layer. For example, the cobbles in a conglomerate must have been formed before the conglomerate.

- The principle of faunal succession states that there is a well-defined order in which organisms have evolved through geological time, and therefore the identification of specific fossils in a rock can be used to determine its age.

In addition to these principles that apply to all sedimentary rocks, a number of other important characteristics of sedimentary processes lead to the development of distinctive sedimentary features in specific sedimentary environments. By understanding the origins of these features, we can make some very useful inferences about the processes that led to deposition the rocks.

# SOFT-SEDIMENT DEFORMATION STRUCTURES

Soft-sediment deformation structures develop at deposition or shortly after, during the first stages of the sediment's consolidation. This is because the sediments need to be "liquid-like" or unsolidified for the deformation to occur. These formations have also been put into a category called water-escape structures by Lowe. The most common places for soft-sediment deformations to materialize are in deep water basins with turbidity currents, rivers, deltas, and shallow-marine areas with storm impacted conditions. This is because these environments have high deposition rates, which allows the sediments to pack loosely.

## Types of Soft-Sediment Deformation Structures

- Convolute bedding forms when complex folding and crumpling of beds or laminations occur. This type of deformation is found in fine or silty sands, and is usually confined to one rock layer. Convolute laminations are found in flood plain, delta, point-bar, and intertidal-flat deposits. They generally range in size from 3 to 25 cm, but there have been larger formations recorded as several meters thick.

- Flame structures consist of mud and are wavy or "flame" shaped. These flames usually extend into an overlying sandstone layer. This deformation is caused from sand being deposited onto mud, which is less dense. Load casts, technically a subset of sole markings,. Flames are thin fingers of mud injected upward into the overlying sands, while load casts are the pendulous knobs of sand that descend downwards into the mud between the flames.

- Slump structures are mainly found in sandy shales and mudstones, but may also be in limestones, sandstones, and evaporites. They are a result of the displacement and movement of unconsolidated sediments, and are found in areas with steep slopes and fast sedimentation rates. These structures often are faulted.

- Dish structures are thin, dish-shaped formations that normally occur in siltstones and sandstones. The size of each "dish" often ranges from 1 cm to 50 cm in size, and forms as a result of dewatering. Pillar structures often appear along with dish structures and also form by dewatering. They have a vertical orientation, which cuts across laminated or massive sands. These formations can range from a few millimeters in diameter to larger than a meter.

- Sole markings are found on the underside of sedimentary rocks that overlie shale beds, usually sandstones. They are used for determining the flow direction of old currents because of their directional features. Sole markings form from the erosion of a bed, which creates a groove that is later filled in by sediment.

- Seismites are sedimentary beds disturbed by seismic waves from earthquakes. They are commonly used to interpret the seismic history of an area. The term has also been applied to soft sediment deformation structures, including sand volcanos, sand blows, and certain clastic dikes.

# BEDDING

The term bedding (also called stratification) ordinarily describes the layering that occurs in sedimentary rocks and sometimes the layering found in metamorphic rock . Bedding may occur when one distinctly different layer of sediment is deposited on an older layer, such as sand and pebbles deposited on silt or when a layer of exposed sedimentary rock has a new layer of sediments deposited on it. Such depositions of sediments produce a clear division between beds called the bedding plane.

The variation among different sedimentary rock layers (usually referred to as beds or strata) may range from subtle to very distinct depending upon color, composition, cementation, texture, or other factors. One of the best examples may be seen in Arizona's Grand Canyon where red, green, white, gray, and other colors heighten the contrast between beds.

The bedding found in metamorphic rock that formed from sedimentary rock is evidence of extreme heat and pressure and is often quite distorted. Distortions may change the sedimentary bedding by compressing, inclining, folding, or other changes.

One of the most common types of bedding is called graded bedding. These beds display a gradual grading from the bottom to the top of the bed with the coarsest sediments at the bottom and the finest at the top. Graded bedding often occurs when a swiftly moving river gradually slows, dropping its heaviest and largest sediments first and lightest last. Changes in a river's speed may be caused by a number of factors, including storm runoff or the entry of a river into a lake or an ocean.

Bedding is usually found in horizontal layers called parallel bedding. But bedding may be inclined or have a swirly appearance. Inclined bedding may occur when sediments are deposited on a slope, such as a sand dune, or when beds are tilted from their original horizontality by forces within the earth. Bedding with a swirly appearance, called cross bedding, may indicate that the sediments making up the rock were deposited by strong desert winds or turbulence in a river.

## Graded Bedding

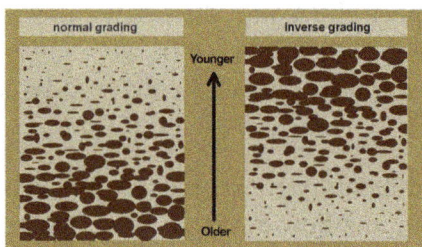

Schematic illustrations of two styles of graded bedding: left: normal grading; right: inverse grading.

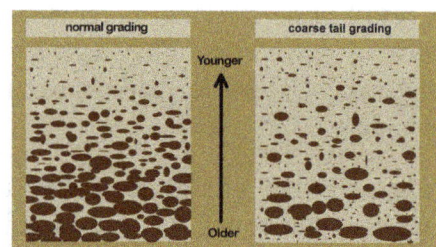

Schematic illustrations of two styles of graded bedding: left: normal grading; right: coarse tail grading.

In geology, a graded bed is one characterized by a systematic change in grain or clast size from one side of the bed to the other. Most commonly this takes the form of normal grading, with coarser sediments at the base, which grade upward into progressively finer ones. Normally graded beds generally represent depositional environments which decrease in transport energy (rate of flow) as time passes, but these beds can also form during rapid depositional events. They are perhaps best represented in turbidite strata, where they indicate a sudden strong current that deposits heavy, coarse sediments first, with finer ones following as the current weakens. They can also form in terrestrial stream deposits.

In reverse or inverse grading the bed coarsens upwards. This type of grading is relatively uncommon but is characteristic of sediments deposited by grain flow and debris flow. It is also observed in Aeolian processes. These deposition processes are examples of granular convection.

## Graded Bedding

Graded bedding is a sorting of particles according to clast size and shape on a lithified horizontal plane. The term is an explanation as to how a geologic profile was formed. Stratification on a lateral plane is the physical result of active depositing of different size materials. Density and gravity forces in the downward movement of these materials in a confined system result in a separating of the detritus settling with respect to size. Thus, finer, higher-porosity clasts form at the top and denser, less porous clasts are consolidated on the bottom, in what is called normal grading. (Inversely graded beds are composed of large clasts on the top, with smaller clasts on the bottom.) Grades of the bedding material are determined by precipitation of solid components compared to the viscosity of the medium in which the particles precipitate. Steno's Principle of Original Horizontality explains that rock layers form in horizontal layers over an underdetermined time scale and depth. Nicholas Steno first published his hypothesis in 1669 after recognizing that fossils were preserved in layers of rock (strata).

## Formation

For materials to settle in stratified layers the defining quality is periodicity. There must be repeated depositional events with changes in precipitation of materials over time. The thickness of graded beds ranges from 1 millimeter to multiple meters. There is no set time limit in which the layers are formed. Uniformity of size and shape of materials within the bed form must be present on a present or previously horizontal plane.

## Necessary Conditions

- Weathering: The chemical or physical forces breaking apart the solid materials that are potentially transported.

- Erosion: The movement of material due to weathering forces that have freed materials for movement.

- Deposition: The material settles on a horizontal plane either through chemical or physical precipitation.

The secondary processes of compaction, cementation, and lithification help to hold a stratified bed in place.

## Sedimentary Graded Bedding

In aeolian or fluid depositional environments, where there is a decrease in transport energy over time, the bedding material is sorted more uniformly, according to the normal grading scale. As water or air slows, the turbidity decreases. The suspended load of the detritus then precipitate. In times of fast movement the bedding may be poorly sorted on the deposition surface and thus is not normally graded because of the quick movement of the material.In broad channels with decreasing slopes, slow-moving water can carry large amounts of detritus over a large area. Thus, graded beds form at points with decreased slopes in wide areas with less bounding of energy current flows. The energy is dispersed and decreases. Turbid sediments precipitate in concordant sizes and shapes in layers.

Changes in currents or physical deformation in the environment can be determined upon observation and monitoring of a depositional surface or lithologic sequence with unconformities above or below a graded bed. Detrital sedimentary graded beds are formed from erosional, depositional, and weathering forces. Graded beds formed from detrital materials are generally composed of sand, and clay. After lithification, shale, siltstone, and sandstone are formed from the detrital deposits.

## Clastic Graded Bedding

Clastic formations are of organic sources, such as biochemical chert, which forms from siliceous marine organism decay and diagenesis. Organic sedimentation of parent material from decaying plant matter in bogs or swamps can also result in a graded bedding complex. This activity leads to formation of peat or coal, after thousands of years. Limestone is more than 95% biogenic in origin. It is made from the deposition of carbonate fossils of marine organisms. Bio erosion caused by animals, such as bivalves, shrimp and sponges change the marine substrate, resulting in layered bedding planes, due to their sifting of bed material in search of food. Organic clastic bedding can become shale and oil shale or millions of years under pressure.

## Cross Bedding

Cross-bedding of sandstone near Mt. Carmel road, Zion Canyon, indicating wind action and sand dune formation had occurred prior to formation of the rock.

In geology, cross-bedding, also known as cross-stratification, is layering within a stratum and at an angle to the main bedding plane. The sedimentary structures which result are roughly horizontal units composed of inclined layers. The original depositional layering is tilted, such tilting not being the result of post-depositional deformation. Cross-beds or "sets" are the groups of inclined layers, which are known as cross-strata.

Cross-bedding forms during deposition on the inclined surfaces of bedforms such as ripples and dunes; it indicates that the depositional environment contained a flowing medium (typically water or wind). Examples of these bedforms are ripples, dunes, anti-dunes, sand waves, hummocks, bars, and delta slopes. Environments in which water movement is fast enough and deep enough to develop large-scale bed forms fall into three natural groupings: rivers, tide-dominated coastal and marine settings.

## Significance

Cross-beds can tell geologists much about what an area was like in ancient times. The direction the beds are dipping indicates paleocurrent, the rough direction of sediment transport. The type and condition of sediments can tell geologists the type of environment (rounding, sorting, composition, etc.) Studying modern analogs allows geologists to draw conclusions about ancient environments. Paleocurrent can be determined by seeing a cross-section of a set of cross-beds. However, to get a true reading, the axis of the beds must be visible. It is also difficult to distinguish between the cross-beds of a dune and the cross-beds of an antidune. (Dunes dip downstream while antidunes dip upstream.)

The direction of motion of the cross-beds can show ancient flow or wind directions (called paleocurrents). The foresets are deposited at the angle of repose (~34 degrees from the horizontal), so geologists are able to measure dip direction of the cross-bedded sediments and calculate the paleoflow direction. However, most cross-beds are not tabular, they are troughs. Since troughs can give a 180 degree variation of the dip of foresets, false paleocurrents can be taken by blindly measuring foresets. In this case, true paleocurrent direction is determined by the axis of the trough. Paleocurrent direction is important in reconstructing past climate and drainage patterns: sand dunes preserve the prevalent wind directions, and current ripples show the direction rivers were moving.

## Formation

Cross-bedding is formed by the downstream migration of bedforms such as ripples or dunes in a flowing fluid. The fluid flow causes sand grains to saltate up the upstream ("stoss") side of the bedform and collect at the peak until the angle of repose is reached. At this point, the crest of granular material has grown too large and will be overcome by the force of the depositing fluid, falling down the downstream ("lee") side of the dune. Repeated avalanches will eventually form the sedimentary structure known as cross-bedding, with the structure dipping in the direction of the paleocurrent.

The sediment that goes on to form cross-stratification is generally sorted before and during deposition on the "lee" side of the dune, allowing cross-strata to be recognized in rocks and sediment deposits.

The angle and direction of cross-beds are generally fairly consistent. Individual cross-beds can range in thickness from just a few tens of centimeters, up to hundreds of feet or more depending upon the depositional environment and the size of the bedform. Cross-bedding can form in any environment in which a fluid flows over a bed with mobile material. It is most common in stream deposits (consisting of sand and gravel), tidal areas, and in aeolian dunes.

## Internal Sorting Patterns

Cross-bedded sediments are recognized in the field by the many layers of "foresets", which are the series of layers that form on the lee side of the bedform (ripple or dune). These foresets are individually differentiable because of small-scale separation between layers of material of different sizes and densities.

Cross-bedding can also be recognized by truncations in sets of ripple foresets, where previously-existing stream deposits are eroded by a later flood, and new bedforms are deposited in the scoured area.

## Geometries

Cross-bedding can be subdivided according to the geometry of the sets and cross-strata into subcategories. The most commonly described types are tabular cross-bedding and trough cross-bedding. Tabular cross-bedding, or planar bedding consists of cross-bedded units that are extensive horizontally relative to the set thickness and that have essentially planar bounding surfaces. Trough cross-bedding, on the other hand, consists of cross-bedded units in which the bounding surfaces are curved, and hence limited in horizontal extent.

## Tabular (Planar) Cross-Beds

Tabular (planar) cross-beds consist of cross-bedded units that are large in horizontal extent relative to set thickness and that have essentially planar bounding surfaces. The foreset laminae of tabular cross-beds are curved so as to become tangential to the basal surface.

Tabular cross-bedding is formed mainly by migration of large-scale, straight-crested ripples and dunes. It forms during lower-flow regimes. Individual beds range in thickness from a few tens of centimeters to a meter or more, but bed thickness down to 10 centimeters has been observed. Where the set height is less than 6 centimeters and the cross-stratification layers are only a few millimeters thick, the term cross-lamination is used, rather than cross-bedding. Cross-bed sets occur typically in granular sediments, especially sandstone, and indicate that sediments were deposited as ripples or dunes, which advanced due to a water or air current.

## Trough Cross-Beds

Cross-beds are layers of sediment that are inclined relative to the base and top of the bed they are associated with. Cross-beds can tell modern geologists many things about ancient environments such as- depositional environment, the direction of sediment transport (paleocurrent) and even environmental conditions at the time of deposition. Typically, units in the rock record are referred to as beds, while the constituent layers that make up the bed are referred to as laminae, when they are less than 1 cm thick and strata when they are greater than 1 cm in thickness. Cross-beds are angled relative to either the base or the top of the surrounding beds. As opposed to angled beds, cross-beds are deposited at an angle rather than deposited horizontally and deformed later on.

Trough cross-beds have lower surfaces which are curved or scoop shaped and truncate the under-lying beds. The foreset beds are also curved and merge tangentially with the lower surface. They are associated with sand dune migration.

## Sediment

The shape of the grains and the sorting and composition of sediment can provide additional infor-mation on the history of cross-beds. Roundness of the grains, limited variation in grain size, and high quartz contents are generally attributed to longer histories of weathering and sediment trans-port. For example: well-rounded, and well-sorted sand that is mostly composed of quartz grains is commonly found in beach environments, far from the source of the sediment. Poorly sorted and angular sediment that is composed of a diversity of minerals is more commonly found in rivers, near the source of the sediment. However, older sedimentary deposits are frequently eroded and re-mobilized. Thus, a river may well erode an older formation of well-rounded, well-sorted beach sands of nearly pure quartz.

## Environments

### Rivers

Flows are characterized by climate (snows, rain, and ice melting) and gradient. Discharge varia-tions measured on a variety of time scales can change water depth, and speed. Some rivers can be characterized by a predictable seasonably controlled hydrograph (reflecting snow melt or rainy season). Others are dominated by durational variations characteristic of alpine glaciers run-off or random storm events, which produce flashy discharge. Few rivers have a long term record of steady flow in the rock record.

Bed forms are relatively dynamic sediment storage bodies with response times that are short rel-ative to major changes in flow characteristics. Large scale bed forms are periodic and occur in the channel (scaled to depth). Their presence and morphologic variability have been related to flow strength expressed as mean velocity or shear stress.

In a fluvial environment, the water in a stream loses energy and its ability transport sediment. The sediment "falls" out of the water and is deposited along a point bar. Over time the river may dry up or avulse and the point bar may be preserved as cross-bedding.

### Tide-Dominated

Tide dominated environments include:

- Coastal water bodies that are partially enclosed by topography, yet have a free connection to the sea.

- Coast lines that have a tidal range of greater than one meter.

- Areas in which the water run-off volume is low relative to the tidal volume or impact.

In general, the greater the tidal range the greater the maximum flow strength. Cross-stratification in tidal-dominated areas can lead to the formation of Herringbone cross-stratification.

Although the flow direction reverses regularly, the flow patterns of flood on ebb currents commonly do not coincide. Consequently, the water and transport sediment may follow a roundabout route in and out of the estuary. This leads to spatially varied systems where some parts of the estuary are flood dominated and other parts are ebb dominated. The temporal and spatial variability of flow and sediment transport, coupled with regular fluctuating water levels creates a variety of bed form morphology.

## Shallow Marine

Large scale bed forms occur on shallow, terrigenous or carbonate clastic continental shelves and epicontinental platforms which are affected by strong geostaphic currents, occasional storm surges and tide currents.

## Aeolian

In an aeolian environment, cross-beds often exhibit inverse grading due to their deposition by grain flows. Winds blow sediment along the ground until they start to accumulate. The side that the accumulation occurs on is called the windward side. As it continues to build, some sediment falls over the end. This side is called the leeward side. Grain flows occur when the windward side accumulates too much sediment, the angle of repose is reached and the sediment tumbles down. As more sediment piles on top the weight causes the underlying sediment to cement together and form cross-beds.

## Bedform

A bedform is a feature that develops at the interface of fluid and a moveable bed, the result of bed material being moved by fluid flow. Examples include ripples and dunes on the bed of a river. Bedforms are often preserved in the rock record as a result of being present in a depositional setting. Bedforms are often characteristic to the flow parameters, and may be used to infer flow depth and velocity, and therefore the Froude number.

Many types of bed forms can be observed in nature. The bed form regimes for steady flow over a sand bed can be classified into:

- Lower transport regime with flat bed, ribbons and ridges, ripples, dunes and bars.

- Transitional regime with washed-out dunes and sand waves.

- Upper transport regime with flat mobile bed and sand waves (anti-dunes).

When the bed form crest is perpendicular (transverse) to the main flow direction, the bed forms are called transverse bed forms, such as ripples, dunes and anti-dunes. Ripples have a length scale much smaller than the water depth, whereas dunes have a length scale much larger than the water depth. The crest lines of the bed forms may be straight, sinuous, linguoid or lunate. Ripples and dunes travel downstream by erosion at the upstream face (stoss-side) and deposition at the downstream face (lee-side). Antidunes travel upstream by lee-side erosion and stoss-side deposition. Bed forms with their crest parallel to the flow are called longitudinal bed forms such as ribbons and ridges.

In the literature, various bed-form classification methods for sand beds are presented. The types of bed forms are described in terms of basic parameters (Froude number, suspension parameter, particle mobility parameter; dimensionless particle diameter).

A flat immobile bed may be observed just before the onset of particle motion, while a flat mobile bed will be present just beyond the onset of motion. The bed surface before the onset of motion may also be covered with relict bed forms generated during stages with larger velocities.

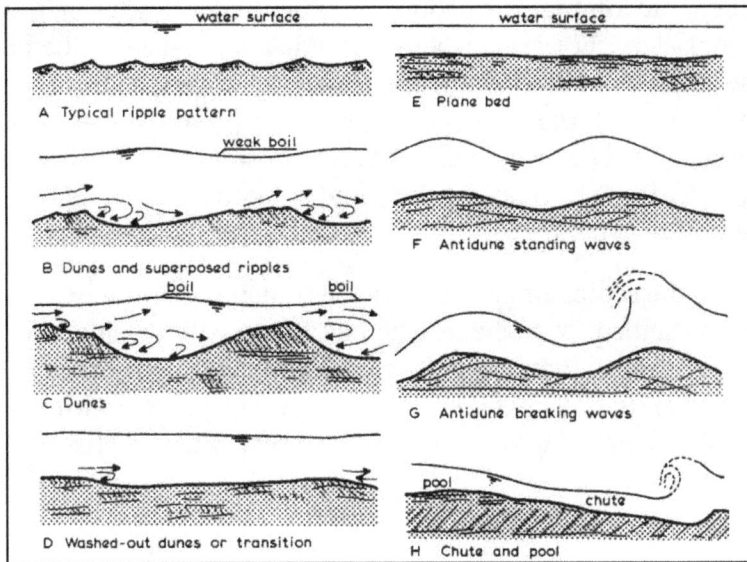

Bed forms in steady flows (rivers).

## Dunes

Another typical bed form type of the lower regime is the dune-type bed form. Dunes have an asymmetrical (triangular) profile with a rather steep lee-side and a gentle stoss-side. A general feature of dune type bed forms is lee-side flow separation resulting in strong eddy motions downstream of the dune crest. The length of the dunes is strongly related to the water depth (h) with values in the range of (3–15)h. Extremely large dunes with heights ($\Delta$) of the order of 7 m and lengths ($\lambda$) of the order of 500 m have been observed in the Rio Parana River (Argentina) at water depths of about 25 m, velocities of about 2 m/s and bed material sizes of about 0.3 mm. The formation of dunes may be caused by large-scale fluid velocity oscillations generating regions at regular intervals with decreased and increased bed-shear stresses, resulting in the local deposition and erosion of sediment particles.

## Sand bars

The largest bed forms in the lower regime are sand bars (such as alternate bars, side bars, point bars, braid bars and transverse bars), which usually are generated in areas with relatively large transverse flow components (bends, confluences, expansions). Alternate bars are features with their crests near alternate banks of the river. Braid bars actually are alluvial "islands" which separate the anabranches of braided streams. Numerous bars can be observed distributed over the cross-sections. These bars have a marked streamwise elongation. Transverse bars are diagonal shoals of triangular-shaped plan along the bed. One side may be attached to the channel bank.

These type of bars generally are generated in steep slope channels with a large width-depth ratio. The flow over transverse bars is sinuous (wavy) in plan. Side bars are bars connected to river banks in a meandering channel. There is no flow over the bar. The planform is roughly triangular. Special examples of side bars are point bars and scroll bars.

## Transitional Regime

It is a well-known phenomenon that the bed forms generated at low velocities are washed out at high velocities. It is not clear, however, whether the disappearance of the bed forms is accomplished by a decrease of the bed form height, by an increase of the bed form length or both. Flume experiments with sediment material of about 0.45 mm show that the transition from the lower to the upper regime is effectuated by an increase of the bed form length and a simultaneous decrease of the bed form height. Ultimately, relatively long and smooth sand waves with a roughness equal to the grain roughness were generated.

In the transition regime the sediment particles will be transported mainly in suspension. This will have a strong effect on the bed form shape. The bed forms will become more symmetrical with relatively gentle lee-side slopes. Flow separation will occur less frequently and the effective bed roughness will approach to that of a plane bed. Large-scale bed forms with a relative height ($\Delta/h$) of 0.1 to 0.2 and a relative length ($\lambda/h$) of 5 to 15 were present in the Mississippi river at high velocities in the upper regime.

## Antidunes

In the supercritical upper regime the bed form types will be plane bed and/or anti-dunes. The latter type of bed forms are sand waves with a nearly symmetrical shape in phase with the water surface waves. The anti-dunes do not exist as a continuous train of bed waves, but they gradually build up locally from a flat bed. Anti-dunes move upstream due to strong lee-side erosion and stoss-side deposition. Anti-dunes are bed forms with a length scale of less than 10 times the water depth. When the flow velocity further increases, finally a stage with chute and pools may be generated.

## Bed Roughness

Nikuradse introduced the concept of an equivalent or effective sand roughness height, ks, to simulate the roughness of arbitrary roughness elements of the bottom boundary. In case of a movable bed consisting of sediments the effective bed roughness ks mainly consists of grain roughness ($k'_s$) generated by skin friction forces and of form roughness ($k''_s$) generated by pressure forces acting on the bed forms. Similarly, a grain-related bed-shear stress ($\tau'b$) and a form-related bed-shear stress ($\tau''_b$) can be defined. The effective bed roughness for a given bed material size is not constant but depends on the flow conditions. Analysis results of ks-values computed from Mississippi River data (USA) show that ks strongly decreases from about 0.5 m at low velocities (0.5 m/s) to about 0.001 m at high velocities (2 m/s), probably because the bed forms become more rounded or are washed out at high velocities. The fundamental problem of bed roughness prediction is that the bed characteristics (bed forms) and hence the bed roughness depend on the main flow variables (depth, velocity) and sediment transport rate (sediment size). These hydraulic variables are, however, in turn strongly dependent on the bed configuration and its roughness. Another problem is the almost continuous variation of the discharge during rising and falling stages. Under these conditions the bed form dimensions and hence the Chézy-coefficient are not constant but vary with the flow conditions.

# RIPPLE MARKS

Ancient wave ripple marks in sandstone, Moenkopi Formation, Capitol Reef National Park, Utah.

In geology, ripple marks are sedimentary structures (i.e., bedforms of the lower flow regime) and indicate agitation by water (current or waves) or wind.

Ripple marks in Cretaceous Dakota Formation, east side of Dinosaur Ridge. Scale bar on notebook is 10 cm.

Ripple beds in the Wren's Nest National Nature Reserve, Dudley, England.

Current ripple marks, unidirectional ripples, or asymmetrical ripple marks are asymmetrical in profile, with a gentle up-current slope and a steeper down-current slope. The down-current slope is the angle of repose, which depends on the shape of the sediment. These commonly form in fluvial and aeolian depositional environments, and are a signifier of the lower part of the Lower Flow Regime.

Ripple cross-laminae forms when deposition takes place during migration of current or wave ripples. A series of cross-laminae are produced by superimposing migrating ripples. The ripples form lateral to one another, such that the crests of vertically succeeding laminae are out of phase and appear to be advancing upslope. This process results in cross-bedded units that have the general appearance of waves in outcrop sections cut normal to the wave *crests*. In sections with other orientations, the laminae may appear horizontal or *trough*-shaped, depending upon the orientation and the shape of the ripples. Ripple cross-laminae will always have a steeper dip downstream, and will always be perpendicular to paleoflow meaning the orientation of the ripples will be in a direction that is ninety degrees to the direction that current if flowing. Scientists suggest current drag, or the slowing of current velocity, during deposition is responsible for ripple cross-laminae.

Wave/symmetrical ripple, Nomgon, Mongolia.

Ripples climb when sediment fluxes in the flow are very high.

## Types

## Straight

Straight ripples generate cross-laminae that all dip in the same direction, and lay in the same plane. These forms of ripples are constructed by unidirectional flow of the current.

## Sinuous

Sinuous ripples generate cross-laminae that are curvy. They show a pattern of curving up and down as shown in picture. Sinuous ripples produce trough cross lamination. All laminae formed under this type of ripple dip at an angle to the flow as well as downstream. These laminae are also formed by the unidirectional current.

## Catenary

Catenary ripples generate cross-laminae that are curvy but have a unidirectional swoop. They show a pattern similar to what a repeated "W" would look like. Like the sinuous ripples, this form of ripple is created by unidirectional flow with the dip at an angle to the flow as well as downstream.

## Linguoid/Lunate

Linguoid ripples have lee slope surfaces that are curved generating a laminae similar to caternary and sinuous ripples. Linguoid ripples generate an angle to the flow as well as downstream. Linguoid ripples have a random shape rather than a "W" shape, as described in the catenary description. Lunate ripples, meaning crescent shaped ripples, are exactly like linguoid ripples except that the stoss sides are curved rather than the lee slope. All other features are the same.

| Size (scale) | Description |
|---|---|
| Very small | Very small cross-lamination means that the ripple height is roughly one centimeter. It is lenticular, wavy and flaser lamination. |
| Small | Small cross-bedding are ripples set at a height less than ten centimeters, while the thickness is only a few millimeters. Some ripples that may fit this category are wind ripples, wave ripples, and current ripples. |
| Medium | Medium cross-lamination are ripples with a height greater than ten centimeters, and less than one meter in thickness. Some ripples that may fit this category would be current-formed sand waves, and storm-generated hummocky cross stratification. |
| Large | Large cross-bedding are ripples with a height greater than one meter, and a thickness equivalent to one meter or greater. Some ripples that may fit this category would be high energy river-bed bars, sand waves, epsilon cross-bedding and Gilbert-type cross-bedding. |

## Ripple Marks in Different Environments

## Wave-Formed Ripples

Tidal megaripples in the Random Formation in Newfoundland.

- Also called bidirectional ripples, or symmetrical ripple marks have a symmetrical, almost sinusoidal profile; they indicate an environment with weak currents where water motion is dominated by wave oscillations.

- In most present-day streams, ripples will not form in sediment larger than coarse sand. Therefore, the stream beds of sand-bed streams are dominated by current ripples, while gravel-bed streams do not contain bedforms. The internal structure of ripples is a base of fine sand with coarse grains deposited on top since the size distribution of sand grains correlates to the size of the ripples. This occurs because the fine grains continue to move while the coarse grains accumulate and provide a protective barrier.

## Ripple Marks Formed by Aeolian Processes

## Normal Ripples

Also known as impact ripples, these occur in the lower part of the lower flow regime sands with grain sizes between 0.3-2.5 mm and normal ripples form wavelengths of 7-14 cm. Normal ripples have straight or slightly sinuous crests approximately transverse to the direction of the wind.

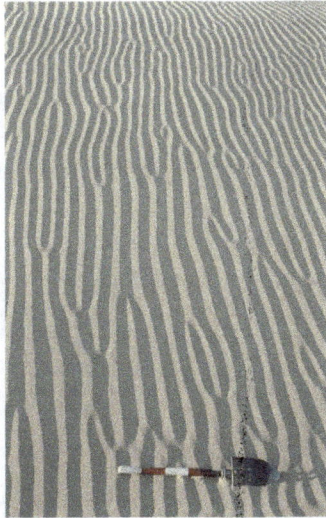

Wind ripples on crescent-shaped sand dunes (Barchans) in Southwest Afghanistan (Sistan).

## Megaripples

These occur in the upper part of the lower flow regime where sand with bimodal particle size distribution forms unusually long wavelength of 1-25 m where the wind is not strong enough to move the larger particles but strong enough to move the smaller grains by saltation.

## Fluid Drag Ripples

Also known as aerodynamic ripples, these are formed with fine, well-sorted grain particles accompanied by high velocity winds which result in long, flat ripples. The flat ripples are formed by long saltation paths taken by grains in suspension and grains on the ground surface.

# MUD CRACKS

Fresh mudcracks.

Mudcracks (also known as desiccation cracks, mud cracks or cracked mud) are sedimentary structures formed as muddy sediment dries and contracts. Crack formation also occurs in clay-bearing soils as a result of a reduction in water content.

Crack pattern in clay exposed to the air.

Naturally forming mudcracks start as wet, muddy sediment dries up and contracts. A strain is developed because the top layer shrinks while the material below stays the same size. When this strain becomes large enough, channel cracks form in the dried-up surface to relieve the strain. Individual cracks spread and join up, forming a polygonal, interconnected network. These cracks may later be filled with sediment and form casts over the base.

Syneresis cracks are broadly similar features that form from underwater shrinkage of muddy sediment caused by differences in salinity or chemical conditions, rather than aerial exposure and desiccation. Syneresis cracks can be distinguished from mudcracks because they tend to be discontinuous, sinuous, and trilete or spindle-shaped.

## Morphology and Classification

Mudcracks are generally polygonal when seen from above and v-shaped in cross section. The "v" opens towards the top of the bed and the crack tapers downward. Allen proposed a classification scheme for mudcracks based on their completeness, orientation, shape, and type of infill.

## Completeness

Complete mudcracks form an interconnected tessellating network. The connection of cracks often occurs when individual cracks join together forming a larger continuous crack. *Incomplete mudcracks* are not connected to each other but still form in the same region or location as the other cracks.

## Plan-view Geometry

Orthogonal intersections can have a preferred orientation or may be random. In oriented orthogonal cracks, the cracks are usually complete and bond to one another forming irregular polygonal shapes and often rows of irregular polygons. In random orthogonal cracks, the cracks are incomplete and unoriented therefore they do not connect or make any general shapes. Although they do not make general shapes they are not perfectly geometric. *Non-orthogonal* mudcracks have a geometric pattern. In uncompleted non-orthogonal cracks they form as a single three point star shape that is composed of three cracks. They could also form with more than three cracks but three cracks in commonly considered the minimum. In completed non-orthogonal cracks, they form a very geometric pattern. The pattern resembles small polygonal shaped tiles in a repetitive pattern.

## Mud Curls

Mud curls form during one of the final stages in desiccation. Mud curls commonly occur on the exposed top layer of very thinly bedded mud rocks. When mud curls form, the water that is inside the sediment begins to evaporate causing the stratified layers to separate. The individual top layer is much weaker than multiple layers and is therefore able to contract and form curls as desiccation occurs. If transported by later currents, mud curls may be preserved as mud-chip rip-up clasts.

## Environments and Substrates

Naturally occurring mudcracks form in sediment that was once saturated with water. Abandoned river channels, floodplain muds, and dried ponds are localities that form mudcracks. Mudcracks can also be indicative of a predominately sunny or shady environment of formation. Rapid drying, which occurs in sunny environments, results in widely spaced, irregular mudcracks, while closer spaced, more regular mudcracks indicate that they were formed in a shady place. Similar features also occur in frozen ground, lava flows (as columnar basalt), and igneous dykes and sills.

## Technology

Polygonal crack networks similar to mudcracks can form in man-made materials such as ceramic glazes, paint film, and poorly made concrete. Mudcrack patterning at smaller scales can also be observed studied using technological thin films deposited using micro and nanotechnologies.

## Preservation

Ancient mudcracks preserved on the base of a bed of sandstone.

Mudcracks can be preserved as v-shaped cracks on the top of a bed of muddy sediment or as casts on the base of the overlying bed. When they are preserved on the top of a bed, the cracks look as they did at the time of formation. When they are preserved on the bottom of the bedrock, the cracks are filled in with younger, overlying sediment. In most bottom-of-bed examples, the cracks are the part that sticks out most. Bottom-of-bed preservation occurs when mudcracks that have already formed and are completely dried are covered with fresh, wet sediment and are buried. Through burial and pressure, the new wet sediment is further pushed into the cracks, where it dries and hardens. The mudcracked rock is then later exposed to erosion. In these cases, the original mud cracks will erode faster than the newer material that fills the spaces. This type of mudcrack is used by geologists to determine the vertical orientation of rock samples that have been altered through folding or faulting.

# DISH STRUCTURE

A dish structure is a type of sedimentary structure formed by liquefaction and fluidization of water-charged soft sediment either during or immediately following deposition. Dish structures are most commonly found in turbidites and other types of clastic deposits that result from subaqueous sediment gravity flows.

Dish structure was described scientifically for the first time by Crook in 1961 who still used the title discontinuous curved lamination. The established term was used for the first time in 1967 by Stauffer and by Wentworth. Comprehensive studies are due to Lowe and LoPiccolo in 1974 and Lowe in 1975.

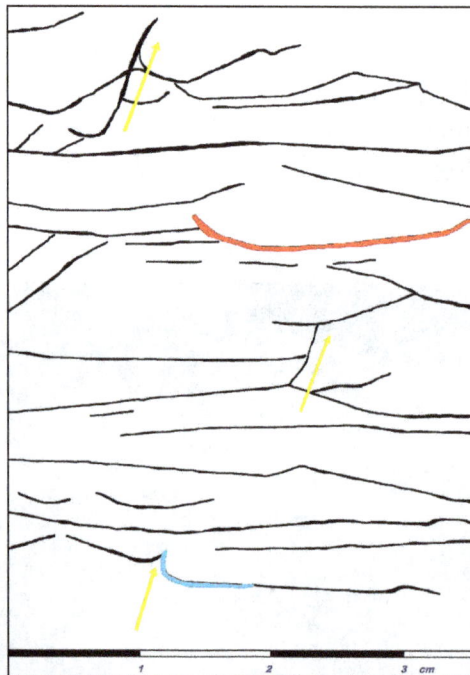

Sketched dish structure from the Jack Fork Group. Nearly perfect dish in red, water escape indicated by yellow arrows. Upturned edge of a dish in blue.

The subhorizontal dish structure consists of two parts, the dish itself and the [sediment] contained within the dish plus the region stretching up to the bounding surface of the overlying dish(or dishes) above. The bounding surface of the dish can take on variable shapes, from substantially flat to bowl-like and to strongly concave up. The bounding surfaces are thin, (and usually) dark(er) laminae; they are richer in clay, silt or organic material than the surrounding sediment. The individual dishes are arranged *en echelon*. Their width can vary from 2 centimeters to over 50 centimeters, the vertical spacing ranges usually from less than 1 centimeter to about 8 centimeters. Their plan shape grades from circular/polygonal to oval/elliptical. Their bases are sharp, but the tops are gradational.

Commonly the dishes are separated by vertical streaks of massive sand called 'pillars'. These pillars can be small-scale structures (Type A pillars) or large and throughgoing high-discharge structures (Type B pillars). Within an individual bed an increase in concavity combined with a simultaneous decrease in width of the dishes can often be observed towards the top.

## Occurrence

Dish structure occurs in laterally extensive sheets. The medium in which the structure forms is usually coarse silt, but it also appears in all grades of sand. They are never found in gravels nor in clays. The containing beds are normally graded. The depositional environment of the structure is mainly deep-water marine (i.e. continental rise) comprising coarser grained turbidity currents and related high-concentration flows (grain flows, fluidized flows, liquefied flows). But dish structure can also be encountered in shallow-marine deposits and in fluviatile, lacustrine and delta environments. It is occasionally found in ash layers within marine sediments.

Giant dish structure near Talara, Peru.

In turbidites dish structure usually forms within Bouma C, occasionally also within Bouma B. Good examples of dish structure can be seen for instance in the Jack Fork Group in Oklahoma, in Ordovician turbidites at Cardigan in Wales, in deep-sea fan deposits near San Sebastián in Spain and in the Cerro Torro Formation of southern Chile. Some of the largest dish structure is found near Talara in northern Peru.

## Formation

Up to 1974 dish structure was still regarded as a primary sedimentary structure. The formation of the structure was thought to be related either to the mechanics of sediment transport or to deposition in high-concentration gravity-flows. Only since Lowe and LoPiccolos's study, the structure is recognized as penecontemporaneous or secondary, formed during the dewatering of rapidly deposited quick or underconsolidated beds.

The postdepositional character of dish structure can sometimes clearly be seen in cut or displaced primary sedimentary structures (like convolute-laminated beds). During the dewatering process less permeable horizons (richer in small grain sizes like dispersed mud) act as barriers to upward flow; the flow is consequently forced sideways until an upward escape is possible. The sideways directed fluid motion has the tendency to leave fines along the low-permeability barriers which eventually become the clay-enriched laminae of the dishes. When the fluid finally finds a possibility to escape vertically it turns up the edges of the dishes. More forceful upward flow creates pillars – which are essentially dewatering pipes.

# PALAEOCHANNEL

A palaeochannel, or paleochannel, is a remnant of an inactive river or stream channel that has been filled or buried by younger sediment. The sediments that the ancient channel is cut into or buried by can be unconsolidated, semi-consolidated, consolidated or lithified.

## Recognition

A palaeochannel is distinct from the overbank deposits of currently-active river channels, including ephemeral water courses that do not regularly flow (such as the Todd River, Central Australia) because the river bed is filled with sedimentary deposits unrelated to the normal bed load of the current drainage pattern.

Many palaeochannels are arranged on old drainage patterns, distinct from the current drainage system of a catchment. For example, palaeochannels may relate to a system of rivers and creeks that drained east-west if the current drainage direction is north-south.

Palaeochannels can be most easily identified as broad erosional channels into a basement that underlies a system of depositional sequences, which may contain several episodes of deposition and represent meandering peneplain streams.

Thereafter, a palaeochannel may form part of the regolith of a region and although it is unconsolidated or partly consolidated, it is currently part of the erosional surface.

Palaeochannels can also be identified according to their age. For example, there are deposits of Tertiary lignites in the Tertiary river systems preserved on top of Archaean basement in the Yilgarn Craton of Western Australia. The river systems have laid in place for 15 to 50 million years and would be considered palaeochannels.

## Formation

Paleochannels form when river channels aggrade and so deposit sediment on their bed. For the channel deposits to be preserved, the flow must not occupy and erode them again. Examples of what may cause long-term preservation include the channels being in a net-depositional environment and being in a subsiding sedimentary basin. Paleochannels may also be preserved in the short-term on non-net-depositional floodplains in which the river migrates or avulses away from its previous course. The preservation is short-term because unless the channel deposits are buried, flow will eventually reoccupy its formerly-occupied course, reworking and eroding the channel deposits.

## Geological Importance

Palaeochannels are important to geology for a number of reasons:

- Understanding movements of faults, which may redirect river systems and so form stranded channels that are, in essence, palaeochannels.

- Preserving Tertiary, Eocene and Holocene sediments and fossils within them, important locations for palaeontology, palaeobotany and archaeology.

- Preserving evidence of older erosional surfaces and levels, which is useful for estimating the net erosional budget of older regolith.

- Preserving sedimentary records, which is useful for understanding climatic conditions, including various isotopic indicators of past rainfall, temperatures and climates, used to understand climate change and global warming.

## Economic Importance

Palaeochannels can host economic ore deposits of uranium, lignite, precious metals such as gold and platinum, heavy minerals such as tin, tungsten, and iron ore preserved as paleo-placer deposits.

# VEGETATION-INDUCED SEDIMENTARY STRUCTURES

Vegetation-induced sedimentary structures (VISS) are primary sedimentary structures formed by the interaction of detrital sediment with in situ plants. VISS provide physical evidence of vegetation's fundamental role in mediating sediment accumulation and erosion in clastic depositional environments. VISS can be broken into seven types, five being hydrodynamic and two being decay-related. The simple hydrodynamic VISS are categorized by centroclinal cross strata, scratch semicircles and upturned beds. The complex hydrodynamic VISS are categorized by coalesced scour fills and scour-and-mound beds. The decay-related VISS are categorized by mudstone-filled hollows and downturned beds.

## Hydrodynamic Structures

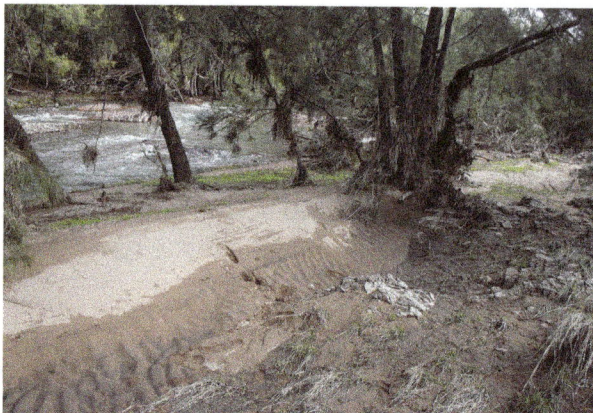

Sediment shadow, Rygel, M.C.

## Upturned Beds (Vegetation Shadow)

Upturned beds are mounds of elongated sediment that are deposited on the lee (downflow) side of an obstruction, containing form-concordant stratification. These may also be described as tongue shaped. These beds are formed where plants were once standing and caused a decrease in bed shear, allowing sediment to deposit near their base. This commonly occurs with meandering flow that deposits sediment on the lee side of a plant. Sediment is also accumulated on the front and sides of an obstruction when deposition is high. The presence of upturned beds in cross-section is consistent with modern vegetation shadows.

## Scratch Circles

Sets of concentric grooves that form in sand or mud by the action of sharp objects anchored at the center of a curvature set. Most likely formed from some sort of free moving plant under water. Plants are bent by the current, causing them to scratch concentric grooves into the adjacent substrate. The grooves are most likely formed in a muddy substrate, which preserved them during deposition of the overlying sediment. These scratch semicircles can record currents in wind or water during ancient deposition.

Scratch circle, Rygel, M.C.

## Centroclinal Cross Strata

Bodies of fine to very fine grained sand that fill symmetrical, cone shaped, depressions centered on tree trunks by swirling flood water. The fill is generally organized into form-concordant, concentric laminae that dip towards the tree. Centroclinal cross strata form from the infilling of scours created around trees by swirling flood water. Studies and piers show us just how the scouring process works. In front of these piers, decelerating flow and friction with the bed material causes a downward pressure gradient that leads to erosive downflow. Downflow is accompanied by horseshoe shaped sediment deposits that generate a U-shaped scour around the front and sides of the obstruction.

Centroclinal-cross-strata, Rygel, M.C.

## Scour-and-Mound Beds

Diffuse sandy lenses associated with standing vegetation at numerous horizons, within the poorly drained floodplain assemblage containing heterolithic bedding. These typically occur above rooted horizons. Scour-and-mound beds that are between sand and heterolithic bedding would suggests that they are formed in forested areas with standing water. Where sand charged flows rapidly decelerated and deposited large amounts of sediment with only minimal scour.

Tree scour, Rygel, M.C.

## Coalesced Scour Fills

Large, internally complex sandstone bodies in well-drained floodplain strata, which superficially resemble small channel bodies. These discrete, locally thickened accumulations are laterally equivalent to thin sheet (crevasse splay) sandstones and are strongly incised into red mudstones. Coalesced scour fills are strongly erosional structures formed where interconnected scours between trees are infilled with sandy sediment during waning flow. Strong incision into underlying strata and downflow tapering suggests that the precursor scours formed in response to vigorous floods across the well-drained floodplain. These essentially are unconstrained by channels and vegetation. The current flow carried only enough sand to fill the scour and blanket the floodplain with a thin sand layer.

## Decay-Related Structures

## Downturned Beds and Mudstone-Filled Hollows

Sedimentary structures that appear to "protrude" into underlying strata. Most likely from the decay of entombed plants. These may have a "pothole-like" form. They reflect a prominent component of soft-sediment deformation in overlying and adjacent strata, but may also represent hydrodynamic activity around a plant that was not preserved.

# HUMMOCKY CROSS-STRATIFICATION

Hummocky cross-stratification is a type of sedimentary structure found in sandstones. It is a form of cross-bedding usually formed by the action of large storms, such as hurricanes. It takes the form of a series of "smile"-like shapes, crosscutting each other. It is only formed at a depth of water below fair-weather wave base and above storm-weather wave base. They are not related to "hummocks" except in shape.

The name was introduced by Harms et al. in 1975. Before this time, these structures were recognized under many different names. When hummocky cross-stratification was founded, it was originally given the name "truncated wave-ripple laminae," by Campbell. The main features were listed by Bourgeois, Harms et al., and Walker, in order to identify the structure. Dott and Bourgeois launched an idealized hummocky stratification sequence. From bottom to top, these include: first-order scoured base (± sole marks); characteristic hummocky zone with several second-order truncation surfaces separating individual undulating lamina sets; a zone of flat laminae; a zone with well-oriented ripple cross-laminae and symmetrical ripple forms; all overlain by a more or less burrowed mudstone or siltstone. Walker wanted to create a second sequence, but it was decided that this sequence offers the best basis for studying hummocky cross-stratification for the future.

## Composition

This structure is commonly found in silt to fine sand. It is typically interbedded with bioturbated mudstone. It commonly contains concretions of abundant mica and plant detritus in the tops of many laminae. This helps indicate a shape sorting. Although hummocky cross-stratification is usually found in shallow marine sedimentary rocks, it has also been found in some lacustrine sedimentary rocks.

## Common Characteristics

Isolated sandy swale within bioturbated mudrocks of the Pebbley Beach Formation
(Permian), New South Wales.

In plan view, it takes on the form of hummocks and swales that are circular to elliptical, with long wavelengths (1–5 m) but with low height (10s of centimeters). Laminations drape these hummocks; in cross-section view, these laminations have an upward curvature, and low angle, curved intersections. Hummocky cross-stratification can form in sediments up to about 3 cm in diameter, with near-bed water particle velocities between about 40–100 cm/s.

## Formation of Structure

This structure is formed under a combination of unidirectional and oscillatory flow that is generated by relatively large storm waves in the ocean. Deposition involves fallout from suspension and lateral tractive flow due to wave oscillation. As the large waves drape sand over an irregular scoured surface, this strong storm-wave action erodes the seabed into low hummocks and swales that lack any significant orientation. It is usually formed by redeposition below normal fair weather wave base delivered offshore by flooding rivers and shoals by large waves.

## Depositional Environments

During ancient times, hummocky cross-stratification was located in shallow marine environments, on the shore face and shelf by waves. It can also form on land during especially large storms when large amounts of water are pushed up onto the tidal flat. These landward deposits feature smaller bed forms due to the attenuation of storm waves as they move onto the land. While it is usually formed in marine settings by the action of storms (e.g.hurricane) it may also be deposited in fluvial strata; a fluvial origin is more likely if the unit solely comprises sand.

# LIESEGANG RINGS

Liesegang rings (also called Liesegangen rings or Liesegang bands) are colored bands of cement observed in sedimentary rocks that typically cut-across bedding. These secondary (diagenetic) sedimentary structures exhibit bands of (authigenic) minerals that are arranged in a regular repeating pattern. Liesegang rings are distinguishable from other sedimentary structures by their concentric or *ring-like* appearance. The precise mechanism from which Liesegang rings form is not entirely known and is still under research, however there is a precipitation process that is thought to be the catalyst for Liesegang ring formation referred to as the Ostwald-Liesegang supersaturation-nucleation-depletion cycle. Though Liesegang rings are considered a frequent occurrence in sedimentary rocks, rings composed of iron oxide can also occur in permeable igneous and metamorphic rocks that have been chemically weathered.

In 1896, a German Chemist named Raphael E. Liesegang first described Liesegang *banding* in his observations from the results of an experiment, and Wilhelm Ostwald provided the earliest explanation for the phenomenon. The purpose of Liesegang's experiment was to observe precipitate formation resulting from the chemical reaction produced when a drop of silver nitrate solution was placed onto the surface of potassium dichromate gel. The resultant precipitate of silver dichromate formed a concentric pattern of rings. Liesegang and successive other workers observed the behavior of precipitates forming rings in sedimentary rocks, hence these features became known as *Liesegang rings*.

## Mechanism for Development

Anvil rock in the Shawnee National Forest, Illinois.

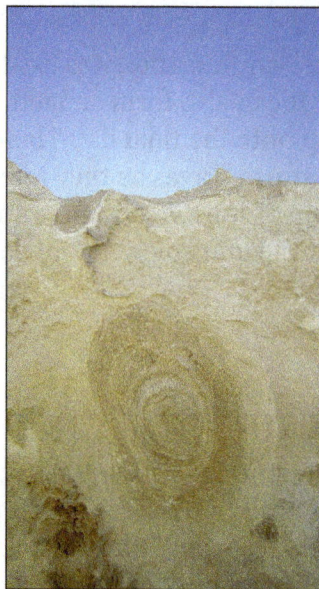

A close view of a Liesegang ring present on a natural arch of sandstone, found at
a beach near Khayelitsha, South Africa.

Liesegang rings (concentric concretions) on east side of Saginaw Hill, Tucson Arizona.

The process by which Liesegang rings develop is not completely understood. Liesegang rings may form from the chemical segregation of iron oxides and other minerals during weathering. One popular mechanism suggested by geochemists is that Liesegang rings develop when there is a lack of convection (advection) and has to do with the inter-diffusion of reacting species such as oxygen and ferrous iron that precipitate in separate discrete bands which become spaced apart in a geometric pattern. A process of precipitation known as the Ostwald-Liesegang supersaturation-nucleation-depletion cycle is known by the geologic community as a probable mechanism for Liesegang ring formation in sedimentary rocks. In this process the "diffusion of reactants leads to supersaturation and nucleation; this precipitation results in localized band formation and depletion of reactants in adjacent zones." As Ostwald suggests, there is a localized formation of crystal seeds that occurs when the right level of supersaturation is reached, and once the crystal seeds form, the growth of the crystals is believed to lower the supersaturation level of fluids in pore spaces surrounding the crystals, thus mineralization that occurs after the initial crystal growth in the surrounding areas develops in bands or *rings*. One classic example based on the Ostwald-Liesegang hypothesis is observed in water and rock interactions where iron hydroxide precipitates in sandstone through pore space.

## Occurrence in the Environment

Liesegang ring patterns are considered to be secondary (diagenetic) sedimentary structures, though they are also found in permeable igneous and metamorphic rocks that have been chemically weathered. Chemical weathering of rocks that leads to the formation of Liesegang rings typically involves the diffusion of oxygen in subterranean water into pore space containing soluble ferrous iron. Liesegang rings usually cut across layers of stratification and occur in many types of rock, some of which more commonly include sandstone and chert. Though there is a high occurrence of Liesegang rings in sedimentary rocks, relatively few scientists have studied their mineralogy and texture in enough detail to write more about them. Liesegang rings are referred to as examples of geochemical self-organization, meaning that their distribution in the rock does not seem to be directly related to features that were established prior to Liesegang ring formation. For instance, in certain types of sedimentary rocks such as carbonate siltstones (calcisiltites), Liesegang ring patterns can be misinterpreted for faults; the rings may appear to be "offset," however the laminae in the rock exhibit an unbroken pattern, therefore the *observed* offset is attributed to pseudofaulting. Pseudofaults are the result of Liesegang rings developing within areas of the rock that are adjacent to each other but at varying stratigraphic levels. Liesegang rings can have the appearance of fine lamination and can be mistaken for laminae when parallel or subparallel to the bedding plane, and are more easily differentiated from laminae when the rings are observed cutting across beds or lamination.

# SOLE MARKINGS

Flute cast from the Book Cliffs of Utah.

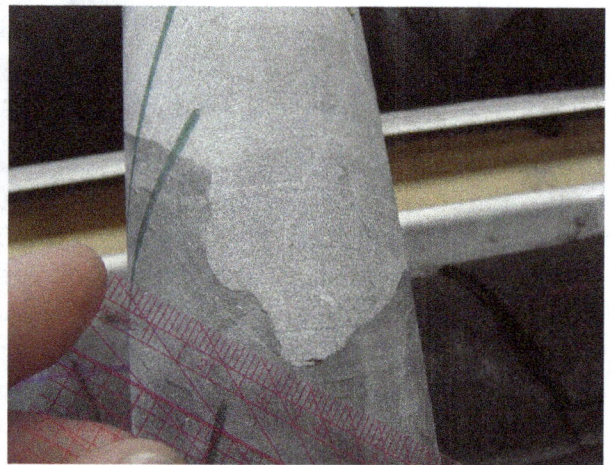

Load cast from drill core.

Sole marks are sedimentary structures found on the bases of certain strata, that indicate small-scale (usually on the order of centimetres) grooves or irregularities. This usually occurs at the interface of two differing lithologies and/or grain sizes. They are commonly preserved as casts of these indents on the bottom of the overlying bed (like flute casts). This is similar to casts and molds in fossil preservation. Occurring as they do only at the bottom of beds, and their distinctive shapes, they can make useful way up structures and paleocurrent indicators.

Sole markings are found most commonly in turbidite deposits, but are also often seen in modern river beds and tidal channels.

## Scour Marks and Flute Casts

Scour marks and flute casts are scours dug into soft, fine sediment which typically get filled by an overlying bed (hence the name cast). Measuring the long axis of the flute cast gives the direction of flow, with the tapered end pointing toward the flow and the steep end up current. The concavity of the flute cast also points stratigraphically up. Flute casts can be characterized into four types, parabolic, spindle-shaped, comet-shaped, and asymmetrical.

- Parabolic flute casts are the most common and simple form. The shape of the bulbous end is parabolic or rounded shape and rarely shows any asymmetrical behavior. They occur in groups or individually, and all move parallel to each other and to paleoflow. In any given environment, their width and length will be consistent and range from a few centimeters, up to one meter in length.

- Spindle-shaped flute casts are found singularly or in groups, are considerably longer than they are wide and have pointed bulbous ends and are generally five to fifteen centimeters long. They can be quite shallow or as much as two thirds of the width in depth. These structures are easily definable because they lack symmetry parallel to the flow direction.

- Comet-shaped flute casts are characteristically found in isolation, and have a sharply pointed bulbous end, but the shallower end shows no stable continuous path. The overall length of the comet-shaped flute is rarely longer than ten centimeters, and the imprints are generally shallow.

- Asymmetrical flute casts are formed on top of a neighboring cast, and, therefore, covers half or more of the underlying flute. As the flutes continue to build outward in a step-like fashion and cut into each other, they get smaller and shallower.

## Tool Marks

Tool marks are a type of sole marking formed by grooves left in a bed by things like sticks being dragged along by a current. The average direction of these can be assumed to be the flow direction, though it is bimodal, so it could be either way along the mark. Tool marks also have a more specific breakdown. There are grooves and striations, skip or prod marks, and roll marks. Groove or striation marks result from the continuous contact with the muddy bed. Skip or prod marks come from objects that bounce along the surface of the muddy bed. And roll marks result from objects rolling along the muddy bed.

- A skip mark is part of a series of linear tool marks left by an object that skipped along the bottom of a stream by saltation. Skip marks are characterized by their even spacing and the crescent-shaped mark that is left on the bed. The skip marks run parallel to paleoflow.

- Saltation is a method of sediment transport that briefly suspends particles and then drops them creating a forward bouncing pattern. This occurs because the turbulent currents are not strong enough to maintain suspension of the particles, but strong enough to suspend the particle for short bursts of time before the particle is returned to the sediment surface and bounces off again.

- A prod mark is a relatively short tool mark caused by an object that was dug into the muddy sediment and then lifted out. These markings are generally asymmetrical, getting deeper down current, and end suddenly.

- Roll marks are made by an object that was forced to roll down the bottom of a stream. The marks made in this case are continuous, long, generally linear, and run parallel to the paleoflow. The width dimensions of roll marks vary based upon the size of the object. Roll marks are a sign of water that has enough energy to cause motion but not enough turbidity and energy to separate the object from the bottom of a muddy bed.

## Groove Casts

Groove casts are straight parallel ridges that are raised a few millimeters from the bedding surface. These structures were named and defined by Shrock in 1948 because of their long and narrow appearance, and they were formed from the filling in of grooves. Even though they may seem similar to flute casts, they each have many distinguishing characteristics, and the two are generally not found in the same vicinity. Groove casts are closely spaced, but not on top of each other, and exist in pairs, triples, and even larger groups. Groove casts form when high velocity flows (e.g. turbidite) create a pattern on an underlying bed. In 1957, Kuenen published that "groove cast" was a general term encompassing both drag marks and slide marks.

- A slide mark is a long, relatively wide, but shallow gouge left in a muddy bed caused by sliding of a soft-body object such as a bed of algae or slumping of sediment.

- Drag marks are narrower and deeper than slide marks, but retain the same length. Drag marks create a groove or striation caused by a physically hard object like a rock or shell.

## Load Casts

Load casts are secondary structures that are preserved as bulbous depressions on the base of a bed. They form as dense, overlying sediment (usually sand) settles into less dense, water-saturated sediment (usually mud) below.

# SEDIMENTARY ROCKS

All rocks, be it igneous, metamorphic, or the already existing sedimentary rocks are constantly subjected to weathering and erosion. Tiny debris from the rock masses and mountains are eroded together with soils, sand, and other granite pieces are normally washed from highlands to low areas. After many years, these materials finally settle down through the process of sedimentation. Some may accumulate under water and others on the lower areas of the land.

The weathering and erosion is normally due to the forces of water, thermal expansion, gravity, wind, and salt crystal expansion that breaks down the pre-existing rocks into small pieces and then carried away to some distance low areas. As the materials move, they are smoothened and rounded by abrasion, and they settle down by leaving pore spaces between the grains which make them achieve their distorted shape.

At this point, they are deposited a layer after layer to form a new sheet of homogenous material. From here, the compaction and cementing agents such as oxides, carbonates, and silica combine together with the deposited material.

The compaction effect due to the weight of the piling layers of materials reduces the porosity of the rocks formed and intensifies the cohesion between the grains. At times, fossil fuels and organic matter may settle within the sediments leading to cementation. Cementation is the gluing of the rock pieces together either by salt compounds or organic matter. When these materials eventually harden, the mixture is transformed into a rock.

Thus, sedimentary rocks are formed from sediment deposits through the process of weathering, erosion, deposition and finally compaction and cementation. Examples of sedimentary rocks include mudstone, limestone, sandstone, and conglomerate.

## Examples of Sedimentary Rocks

As noted from the previous discussions, there are several types of sedimentary rocks. Here are the detailed examples of the various sedimentary rocks.

- Breccia: Brecia are clastic sedimentary rocks made up of angular rock broken parts that are cemented together. The angular shape means that the broken parts haven't traveled far from their pre-existing materials. The broken pieces are similar to conglomerate because of their large pea-sizes. Breccias are commonly found along fault zones and they take any color.

- Conglomerate: Conglomerates are clastic sedimentary rocks composed of semi-rounded rock fragments cemented together. The rounded fragments depict that they have undergone abrasion and traveled a significant distant from their pre-existing materials. Conglomerates fragments are commonly deposited along the shoreline or stream channel and they are pea-sized or larger. They are also sometimes referred to as pudding stone.

- Siltstone: Siltstones are composed of small-sized rock particles which are finer than sand grains but coarser than clay. It is among the clastic sedimentary rocks which are the most difficult to identify since it appears almost similar to fine-grained sandstone or a coarse shale. They normally occur in a wide variety of colors.

- Sandstone: Sandstones are clastic sedimentary rocks made up of cemented sand grains. Sandstones vary from fine-grained to coarse grained are readily distinguishable by the naked eyes. Mature sandstones or quartz sandstones are light-colored and majorly consist of rounded and well-sorted quartz grains. Graywackes or immature sandstones consist of angular grains of diverse minerals. Sandstones are generally white, red, gray, pink, black, or brown in color.

- Shale: Shale consists of clay minerals or clay-sized pieces that have been compacted by the weight of the overlying rock materials. Shale belongs to clastic sedimentary rocks and they tend to split into fairly flat pieces. Shales are of many colors including gray, red, brown, or black depending on their composition of iron oxides and organic materials.They are generally a good source of fossils and are mostly found at the bottom of lakes or oceans.

- Dolomite: Dolomites are chemical sedimentary rocks that almost resemble calcite. The similarity is because dolomite at first begins to take shape as limestone but they are later chemically altered through the substitution of some of its calcium by magnesium.

- Chert: Cherts are chemical sedimentary rocks formed due to the deposition of cryptocrystalline quartz. Cherts are of dull brown or gray in color and are often found as nodules firmly enclosed in limestone which protrude out of the limestone when the limestone is slowly immersed in water. Jasper is red, bright yellowish brown or reddish brown chert. Flint on the other hand is chert with a waxy luster.

- Limestone: Limestones are chemical sedimentary rocks made up of the mineral calcite. They may be hard to visually identify. Their colors vary from brown, dark gray, to light gray. The common types of limestone include fossiliferrous limestone rich in fossils, lithographic limestone that is very fine-grained, coquina limestone composed of broken shell fragments, encrinal limestone composed of crinoid fragments, and travertine deposited by the forces of moving surface water.

- Rock salt: Rock salt is chemical sedimentary rocks often made up of the mineral sodium chloride.Salt is colorless or white and might be colored when mixed with impurities such as clay or iron oxide. It is easily identified by its salty taste and it is also water soluble. It also has the mineral name 'halite.'

- Gypsum: Gypsum belongs to chemical sedimentary rocks. It is soft and can be easily bruised. It is usually white in color and is used to produce plaster of Paris.

- Amber: Amber is an organic sedimentary rock and is naturally plastic and is light-weight compared to majority of the typical stones. Amber is simply a hardened tree sap and its colors ranges from transparent yellow to creamy yellow or red to dark brown.

- Coal: Coal belongs to organic sedimentary rocks are made up from the buildup of decomposed plant material in a swampy environment. Coal is combustible in nature and is frequently extracted for use as fuel. It ranges from brown to black in color and its concentration depends on the compaction and alterations of the pre-existing organic materials. Examples of coal in order of the degree of compaction and alteration include peat, lignite, bituminous, and anthracite.

## CLASTIC SEDIMENTARY ROCKS

A clast is a fragment of rock or mineral, ranging in size from less than a micron (too small to see) to as big as an apartment block. Most sand-sized clasts are made of quartz because quartz is more resistant to weathering than any other common mineral. Most clasts that are smaller than sand size (<1/16 mm) are made of clay minerals. Most clasts larger than sand size (>2 mm) are actual fragments of rock, and commonly these might be fine-grained rock like basalt or andesite, or if they are bigger, coarse-grained rock like granite or gneiss.

There are six main grain-size categories; five are broken down into subcategories, with clay being the exception. The diameter limits for each successive subcategory are twice as large as the one beneath it. In general, a boulder is bigger than a toaster and difficult to lift. There is no upper limit to the size of boulder. A small cobble will fit in one hand, a large one in two hands. A pebble is something that you could throw quite easily. The smaller ones — known as granules — are gravel size, but still you could throw one. But you can't really throw a single grain of sand. Sand ranges from 2 mm down to 0.063 mm, and its key characteristic is that it feels "sandy" or gritty between your fingers — even the finest sand grains feel that way. Silt is essentially too small for individual grains to be visible, and while sand feels sandy to your fingers, silt feels smooth to your fingers but gritty in your mouth. Clay is so fine that it feels smooth even in your mouth.

If you drop a granule into a glass of water, it will sink quickly to the bottom (less than half a second). If you drop a grain of sand into the same glass, it will sink more slowly (a second or two depending on the size). A grain of silt will take several seconds to get to the bottom, and a particle of fine clay may never get there. The rate of settling is determined by the balance between gravity and friction, as shown in figure.

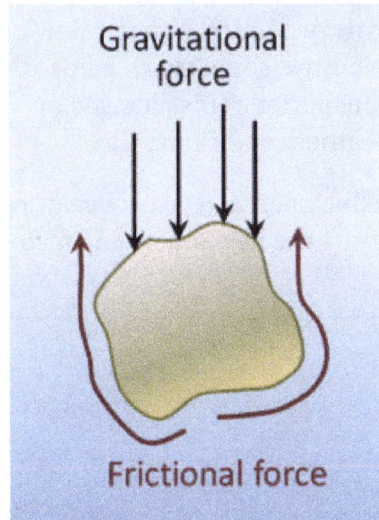

The two forces operating on a grain of sand in water. Gravity is pushing it down, and the friction between the grain and the water is resisting that downward force. Large particles settle quickly because the gravitational force (which is proportional to the mass, and therefore to the volume of the particle) is much greater than the frictional force (which is proportional to the surface area of the particle). For small particles it is only slightly greater, so they settle slowly.

## Transportation

One of the key principles of sedimentary geology is that the ability of a moving medium (air or water) to move sedimentary particles, and keep them moving, is dependent on the velocity of flow. The faster the medium flows, the larger the particles it can move.

Variations in flow velocity on the Englishman River near Parksville, B.C.

Parts of the river are moving faster than other parts, especially where the slope is greatest and the channel is narrow. Not only does the velocity of a river change from place to place, but it changes from season to season.

During peak discharge at this location, the water is high enough to flow over the embankment on the right, and it flows fast enough to move the boulders that cannot be moved during low flows.

Clasts within streams are moved in several different ways, as illustrated in figure. Large bedload clasts are pushed (by traction) or bounced along the bottom (saltation), while smaller clasts are suspended in the water and kept there by the turbulence of the flow. As the flow velocity changes, different-sized clasts may be either incorporated into the flow or deposited on the bottom. At various places along a river, there are always some clasts being deposited, some staying where they are, and some being eroded and transported. This changes over time as the discharge of the river changes in response to changing weather conditions.

Other sediment transportation media, such as waves, ocean currents, and wind, operate under similar principles, with flow velocity as the key underlying factor that controls transportation and deposition.

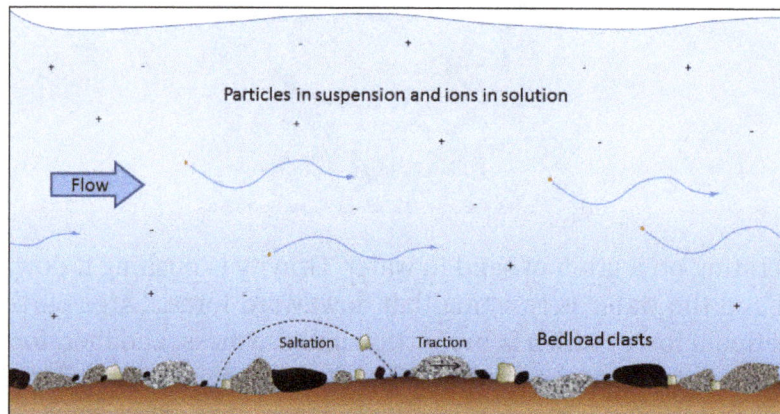

Transportation of sediment clasts by stream flow. The larger clasts, resting on the bottom (bedload), are moved by traction (sliding) or by saltation (bouncing). Smaller clasts are kept in suspension by turbulence in the flow. Ions (depicted as + and − in the image, but invisible in real life) are dissolved in the water.

Clastic sediments are deposited in a wide range of environments, including glaciers, slope failures, rivers — both fast and slow, lakes, deltas, and ocean environments — both shallow and deep. Depending on the grain size in particular, they may eventually form into rocks ranging from fine mudstone to coarse breccia and conglomerate.

Lithification is the term used to describe a number of different processes that take place within a deposit of sediment to turn it into solid rock. One of these processes is burial by other sediments, which leads to compaction of the material and removal of some of the intervening water and air. After this stage, the individual clasts are all touching one another. Cementation is the process of crystallization of minerals within the pores between the small clasts, and also at the points of contact between the larger clasts (sand size and larger). Depending on the pressure, temperature, and chemical conditions, these crystals might include calcite, hematite, quartz, clay minerals, or a range of other minerals.

The characteristics and distinguishing features of clastic sedimentary rocks are summarized in Table. Mudrock is composed of at least 75% silt- and clay-sized fragments. If it is dominated by clay, it is called claystone. If it shows evidence of bedding or fine laminations, it is shale; otherwise it is mudstone. Mudrocks form in very low energy environments, such as lakes, river backwaters, and the deep ocean.

Table: The Main Types of Clastic Sedimentary Rocks and their Characteristics.

| Group | Examples | Characteristics |
|---|---|---|
| Mudrock | mudstone | >75% silt and clay, not bedded |
| | shale | >75% silt and clay, thinly bedded |
| Coal | | dominated by fragments of partially decayed plant matter, often enclosed between beds of sandstone or mudrock |
| Sandstone | quartz sandstone | dominated by sand, >90% quartz |
| | arkose | dominated by sand, >10% feldspar |
| | lithic wacke | dominated by sand, >10% rock fragments, >15% silt and clay |
| Conglomerate | | dominated by rounded clasts, pebble size and larger |
| Breccia | | dominated by angular clasts, pebble size and larger |

Most coal forms in fluvial or delta environments where vegetation growth is vigorous and where decaying plant matter accumulates in long-lasting swamps with low oxygen levels. To avoid oxidation and breakdown, the organic matter must remain submerged for centuries or millennia, until it is covered with another layer of either muddy or sandy sediments.

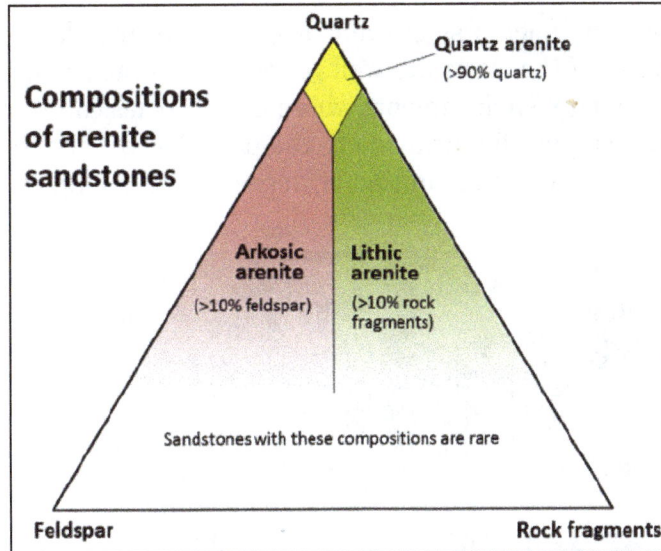

A compositional triangle for arenite sandstones, with the three most common components of sand-sized grains: quartz, feldspar, and rock fragments. Arenites have less than 15% silt or clay. Sandstones with more than 15% silt and clay are called wackes (e.g., quartz wacke, lithic wacke).

It is important to note that in some textbooks coal is described as an "organic sedimentary rock." In this book, coal is classified with the clastic rocks for two reasons: first, because it is made up of fragments of organic matter; and second, because coal seams (sedimentary layers) are almost always interbedded with layers of clastic rocks, such as mudrock or sandstone. In other words, coal accumulates in environments where other clastic rocks accumulate.

It's worth taking a closer look at the different types of sandstone because sandstone is a common and important sedimentary rock. Typical sandstone compositions are shown in figure. The term arenite applies to a so-called clean sandstone, meaning one with less than 15% silt and clay. Considering the sand-sized grains only, arenites with 90% or more quartz are called quartz arenites. If they have more than 10% feldspar and more feldspar than rock fragments, they are called feldspathic arenites or arkosic arenites (or just arkose). If they have more than 10% rock fragments, and more rock fragments than feldspar, they are lithic arenites. A sandstone with more than 15% silt or clay is called a wacke. The terms quartz wacke, lithic wacke, and feldspathic wacke are used. Another name for a lithic wacke is greywacke.

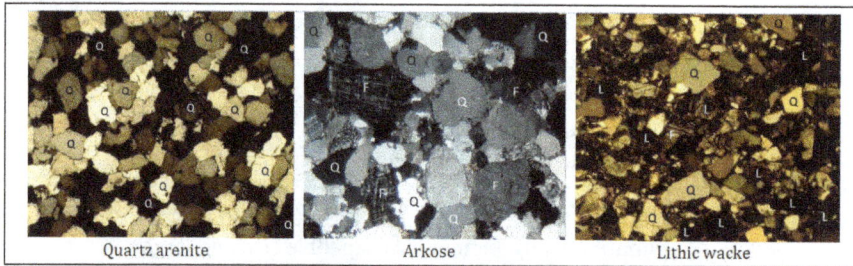

Photos of thin sections of three types of sandstone. Some of the minerals are labelled: Q=quartz, F=feldspar and L= lithic (rock fragments). The quartz arenite and arkose have relatively little silt-clay matrix, while the lithic wacke has abundant matrix.

Some examples of sandstones, magnified in thin section are shown in figure. (A thin section is rock sliced thin enough so that light can shine through.)

Clastic sedimentary rocks in which a significant proportion of the clasts are larger than 2 mm are known as conglomerate if the clasts are well rounded, and breccia if they are angular. Conglomerates form in high-energy environments where the particles can become rounded, such as fast-flowing rivers. Breccias typically form where the particles are not transported a significant distance in water, such as alluvial fans and talus slopes. Some examples of clastic sedimentary rocks are shown on figure.

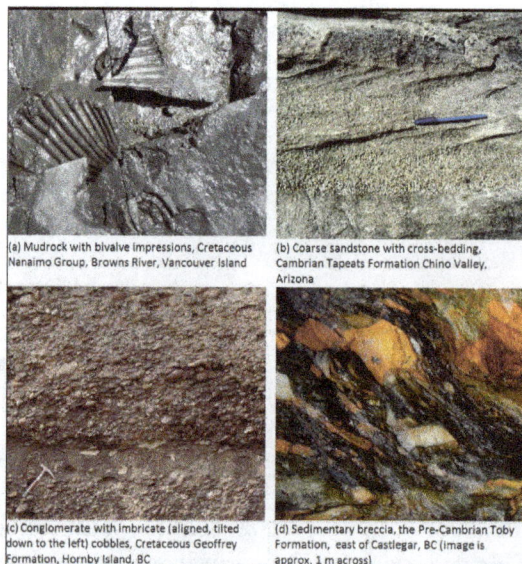

Examples of various clastic sedimentary rocks.

# CONGLOMERATE

Conglomerate is a coarse-grained clastic sedimentary rock that is composed of a substantial fraction of rounded to subangular gravel-size clasts, e.g., granules, pebbles, cobbles, and boulders, larger than 2 mm (0.079 in) in diameter. Conglomerates form by the consolidation and lithification of gravel. Conglomerates typically contain finer grained sediment, e.g., either sand, silt, clay or combination of them, called *matrix* by geologists, filling their interstices and are often cemented by calcium carbonate, iron oxide, silica, or hardened clay.

The size and composition of the gravel-size fraction of a conglomerate may or may not vary in composition, sorting, and size. In some conglomerates, the gravel-size class consist almost entirely of what were clay clasts at the time of deposition. Conglomerates can be found in sedimentary rock sequences of all ages but probably make up less than 1 percent by weight of all sedimentary rocks. In terms of origin and depositional mechanisms, they are closely related to sandstones and exhibit many of the same types of sedimentary structures, e.g., tabular and trough cross-bedding and graded bedding.

## Classification of Conglomerates

Conglomerates may be named and classified by the:

- Amount and type of matrix present.
- Composition of gravel-size clasts they contain.
- Size range of gravel-size clasts present.

A sedimentary rock composed largely of gravel is first named according to the roundness of the gravel. If the gravel clasts that comprise it is largely well-rounded to subrounded, it is a *conglomerate*. If the gravel clasts that comprise it are largely angular, it is a breccia. Such breccias can be called sedimentary breccias to differentiate them from other types of breccia, e.g. volcanic and fault breccias. Sedimentary rocks that contain a mixture of rounded and angular gravel clasts are sometimes called breccio-conglomerate.

## Texture

Conglomerates are rarely composed entirely of gravel-size clasts. Typically, the space between the gravel-size clasts is filled by a mixture composed of varying amounts of silt, sand, and clay, known as *matrix*. If the individual gravel clasts in a conglomerate are separated from each other by an abundance of matrix such that they are not in contact with each other and *float* within the matrix, it is called a paraconglomerate. Paraconglomerates are also often unstratified and can contain more matrix than gravel clasts. If the gravel clasts of a conglomerate are in contact with each other, it is called an orthoconglomerate. Unlike paraconglomerates, orthoconglomerates are typically cross-bedded and often well-cemented and lithified by either calcite, hematite, quartz, or clay.

The differences between paraconglomerates and orthoconglomerates reflect differences in how they are deposited. Paraconglomerates are commonly either glacial tills or debris flow deposits. Orthoconglomerates are typically associated with aqueous currents.

A conglomerate at the base of the Cambrian in the Black Hills, South Dakota.

Section of polymict conglomerate from offshore rock core, Alaska, approximate depth 10,000 ft.

## Clast Composition

Conglomerates are also classified according to the composition of their clasts. A conglomerate or any clastic sedimentary rock that consists of a single rock or mineral is known as either a monomict, monomictic, oligomict, or oligomictic conglomerate. If the conglomerate consists of two or more different types of rocks, minerals, or combination of both, it is known as either a polymict or polymictic conglomerate. If a polymictic conglomerate contains an assortment of the clasts of metastable and unstable rocks and minerals, it called either a petromict or petromictic conglomerate.

In addition, conglomerates are classified by source as indicated by the lithology of the gravel-size clasts If these clasts consist of rocks and minerals that are significantly different in lithology from the enclosing matrix and, thus, older and derived from outside the basin of deposition, the conglomerate is known as an extraformational conglomerate. If these clasts consist of rocks and minerals that are identical to or consistent with the lithology of the enclosing matrix and, thus, penecontemporaneous and derived from within the basin of deposition, the conglomerate is known as an intraformational conglomerate.

Two recognized types of type of intraformational conglomerates are shale-pebble and flat-pebble conglomerates. A shale-pebble conglomerate is a conglomerate that is composed largely of clasts of rounded mud chips and pebbles held together by clay minerals and created by erosion within environments such as within a river channel or along a lake margin. Flat-pebble conglomerates (edgewise conglomerates) are conglomerates that consist of relatively flat clasts of lime mud created by either storms or tsunami eroding a shallow sea bottom or tidal currents eroding tidal flats along a shoreline.

## Clast Size

Finally, conglomerates are often differentiated and named according to the dominant clast size comprising them. In this classification, a conglomerate composed largely of granule-size clasts would be called a granule conglomerate; a conglomerate composed largely of pebble-size clasts would be called a pebble conglomerate; and a conglomerate composed largely of cobble-size clasts would be called a cobble conglomerate.

## Sedimentary Environments

Conglomerates are deposited in a variety of sedimentary environments.

## Deepwater Marine

In turbidites, the basal part of a bed is typically coarse-grained and sometimes conglomeratic. In this setting, conglomerates are normally very well sorted, well-rounded and often with a strong A-axis type imbrication of the clasts.

## Shallow Marine

Conglomerates are normally present at the base of sequences laid down during marine transgressions above an unconformity, and are known as basal conglomerates. They represent the position of the shoreline at a particular time and are diachronous.

## Fluvial

Conglomerates deposited in fluvial environments are typically well rounded and well sorted. Clasts of this size are carried as bedload and only at times of high flow-rate. The maximum clast size decreases as the clasts are transported further due to attrition, so conglomerates are more characteristic of immature river systems. In the sediments deposited by mature rivers, conglomerates are generally confined to the basal part of a channel fill where they are known as *pebble lags*. Conglomerates deposited in a fluvial environment often have an AB-plane type imbrication.

## Alluvial

Fanglomerate in Death Valley National Park.

Alluvial deposits form in areas of high relief and are typically coarse-grained. At mountain fronts individual alluvial fans merge to form braidplains and these two environments are associated with the thickest deposits of conglomerates. The bulk of conglomerates deposited in this setting are clast-supported with a strong AB-plane imbrication. Matrix-supported conglomerates, as a result of debris-flow deposition, are quite commonly associated with many alluvial fans. When such conglomerates accumulate within an alluvial fan, in rapidly eroding (e.g., desert) environments, the resulting rock unit is often called a *fanglomerate*.

## Glacial

Glaciers carry a lot of coarse-grained material and many glacial deposits are conglomeratic. Til-lites, the sediments deposited directly by a glacier, are typically poorly sorted, matrix-supported conglomerates. The matrix is generally fine-grained, consisting of finely milled rock fragments. Waterlaid deposits associated with glaciers are often conglomeratic, forming structures such as eskers.

## Examples

An example of conglomerate can be seen at Montserrat, near Barcelona. Here, erosion has created vertical channels that give the characteristic jagged shapes the mountain is named for (Montserrat literally means "jagged mountain"). The rock is strong enough to use as a building material, as in the Santa Maria de Montserrat Abbey.

Another example, the Crestone Conglomerate, occurs in and near the town of Crestone, at the foot of the Sangre de Cristo Range in Colorado's San Luis Valley. The Crestone Conglomerate consists of poorly sorted fanglomerates that accumulated in prehistoric alluvial fans and related fluvial systems. Some of these rocks have hues of red and green.

Conglomerate cliffs are found on the east coast of Scotland from Arbroath northwards along the coastlines of the former counties of Angus and Kincardineshire. Dunnottar Castle sits on a rugged promontory of conglomerate jutting into the North Sea just south of the town of Stonehaven.

Conglomerate may also be seen in the domed hills of Kata Tjuta, in Australia's Northern Territory.

In the nineteenth century a thick layer of Pottsville conglomerate was recognized to underlie anthracite coal measures in Pennsylvania.

## Examples on Mars

On Mars, slabs of conglomerate have been found at an outcrop named "Hottah", and have been interpreted by scientists as having formed in an ancient streambed. The gravels, which were discovered by NASA's Mars rover Curiosity, range from the size of sand particles to the size of golf balls. Analysis has shown that the pebbles were deposited by a stream that flowed at walking pace and was ankle- to hip-deep.

# SANDSTONE

Sandstone is a clastic sedimentary rock composed mainly of sand-sized (0.0625 to 2 mm) mineral particles or rock fragments.

Most sandstone is composed of quartz or feldspar (both silicates) because they are the most resistant minerals to weathering processes at the Earth's surface, as seen in Bowen's reaction series. Like uncemented sand, sandstone may be any color due to impurities within the minerals, but the most common colors are tan, brown, yellow, red, grey, pink, white, and black. Since sandstone beds often form highly visible cliffs and other topographic features, certain colors of sandstone have been strongly identified with certain regions.

Rock formations that are primarily composed of sandstone usually allow the percolation of water and other fluids and are porous enough to store large quantities, making them valuable aquifers and petroleum reservoirs. Fine-grained aquifers, such as sandstones, are better able to filter out pollutants from the surface than are rocks with cracks and crevices, such as limestone or other rocks fractured by seismic activity.

Quartz-bearing sandstone can be changed into quartzite through metamorphism, usually related to tectonic compression within orogenic belts.

Sand from Coral Pink Sand Dunes State Park, Utah. These are grains of quartz with a hematite coating providing the orange colour. Scale bar is 1.0 mm.

Sandstones are *clastic* in origin (as opposed to either *organic*, like chalk and coal, or *chemical*, like gypsum and jasper). They are formed from cemented grains that may either be fragments of a pre-existing rock or be mono-minerallic crystals. The cements binding these grains together are typically calcite, clays, and silica. Grain sizes in sands are defined (in geology) within the range of 0.0625 mm to 2 mm (0.0025–0.08 inches). Clays and sediments with smaller grain sizes not visible with the naked eye, including siltstones and shales, are typically called *argillaceous* sediments; rocks with larger grain sizes, including breccias and conglomerates, are termed *rudaceous* sediments.

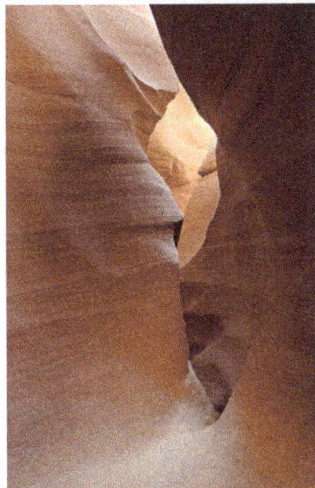

Red sandstone interior of Lower Antelope Canyon, Arizona, worn smooth by erosion from flash flooding over thousands of years.

Tafoni at Twyfelfontein in Namibia.

The formation of sandstone involves two principal stages. First, a layer or layers of sand accumulates as the result of sedimentation, either from water (as in a stream, lake, or sea) or from air (as in a desert). Typically, sedimentation occurs by the sand settling out from suspension; i.e., ceasing to be rolled or bounced along the bottom of a body of water or ground surface (e.g., in a desert or erg). Finally, once it has accumulated, the sand becomes sandstone when it is compacted by the pressure of overlying deposits and cemented by the precipitation of minerals within the pore spaces between sand grains.

The most common cementing materials are silica and calcium carbonate, which are often derived either from dissolution or from alteration of the sand after it was buried. Colors will usually be tan or yellow (from a blend of the clear quartz with the dark amber feldspar content of the sand). A predominant additional colourant in the southwestern United States is iron oxide, which imparts reddish tints ranging from pink to dark red (terracotta), with additional manganese imparting a purplish hue. Red sandstones, both Old Red Sandstone and New Red Sandstone, are also seen in the Southwest and West of Britain, as well as central Europe and Mongolia. The regularity of the latter favours use as a source for masonry, either as a primary building material or as a facing stone, over other forms of construction.

The environment where it is deposited is crucial in determining the characteristics of the resulting sandstone, which, in finer detail, include its grain size, sorting, and composition and, in more general detail, include the rock geometry and sedimentary structures. Principal environments of deposition may be split between terrestrial and marine, as illustrated by the following broad groupings:

- Terrestrial environments:

  ◦ Rivers (levees, point bars, channel sands).

  ◦ Alluvial fans.

  ◦ Glacial outwash.

  ◦ Lakes.

  ◦ Deserts (sand dunes and ergs).

- Marine environments:

  ◦ Deltas.

- ○ Beach and shoreface sands.

- ○ Tidal flats.

- ○ Offshore bars and sand waves.

- ○ Storm deposits (tempestites).

- ○ Turbidites (submarine channels and fans).

## Components

## Framework Grains

Paradise Quarry, Sydney, Australia.

Grus sand and the granitoid from which it is derived.

Framework grains are sand-sized (0.0625-to-2-millimetre (0.00246 to 0.07874 in) diameter) detrital fragments that make up the bulk of a sandstone. These grains can be classified into several different categories based on their mineral composition:

- Quartz framework grains are the dominant minerals in most clastic sedimentary rocks; this is because they have exceptional physical properties, such as hardness and chemical stability. These physical properties allow the quartz grains to survive multiple recycling events, while also allowing the grains to display some degree of rounding. Quartz grains evolve from plutonic rock, which are felsic in origin and also from older sandstones that have been recycled.

- Feldspathic framework grains are commonly the second most abundant mineral in sandstones. Feldspar can be divided into two smaller subdivisions: alkali feldspars and plagioclase feldspars. The different types of feldspar can be distinguished under a petrographic microscope.

  ◦ Alkali feldspar is a group of minerals in which the chemical composition of the mineral can range from $KAlSi_3O_8$ to $NaAlSi_3O_8$, this represents a complete solid solution.

  ◦ Plagioclase feldspar is a complex group of solid solution minerals that range in composition from $NaAlSi_3O_8$ to $CaAl_2Si_2O_8$.

Volcanic sand grain; upper picture is plane-polarised light, bottom picture is cross-polarised light, scale box at left-centre is 0.25 millimetre. This type of grain would be a main component of a lithic sandstone.

- Lithic framework grains are pieces of ancient source rock that have yet to weather away to individual mineral grains, called lithic fragments or clasts. Lithic fragments can be any fine-grained or coarse-grained igneous, metamorphic, or sedimentary rock, although the most common lithic fragments found in sedimentary rocks are clasts of volcanic rocks.

- Accessory minerals are all other mineral grains in a sandstone; commonly these minerals make up just a small percentage of the grains in a sandstone. Common accessory minerals include micas (muscovite and biotite), olivine, pyroxene, and corundum. Many of these accessory grains are more dense than the silicates that make up the bulk of the rock. These heavy minerals are commonly resistant to weathering and can be used as an indicator of sandstone maturity through the ZTR index. Common heavy minerals include zircon, tourmaline, rutile (hence *ZTR*), garnet, magnetite, or other dense, resistant minerals derived from the source rock.

## Matrix

Matrix is very fine material, which is present within interstitial pore space between the framework grains. The nature of the matrix within the interstitial pore space results in a twofold classification:

- Arenites are texturally *clean* sandstones that are free of or have very little matrix.

- Wackes are texturally *dirty* sandstones that have a significant amount of matrix.

## Cement

Cement is what binds the siliciclastic framework grains together. Cement is a secondary mineral that forms after deposition and during burial of the sandstone. These cementing materials may be either silicate minerals or non-silicate minerals, such as calcite.

- Silica cement can consist of either quartz or opal minerals. Quartz is the most common silicate mineral that acts as cement. In sandstone where there is silica cement present, the quartz grains are attached to cement, which creates a rim around the quartz grain called overgrowth. The overgrowth retains the same crystallographic continuity of quartz framework grain that is being cemented. Opal cement is found in sandstones that are rich in volcanogenic materials, and very rarely is in other sandstones.

- Calcite cement is the most common carbonate cement. Calcite cement is an assortment of smaller calcite crystals. The cement adheres itself to the framework grains, this adhesion is what causes the framework grains to be adhered together.

- Other minerals that act as cements include: hematite, limonite, feldspars, anhydrite, gypsum, barite, clay minerals, and zeolite minerals.

## Pore space

Pore space includes the open spaces within a rock or a soil. The pore space in a rock has a direct relationship to the porosity and permeability of the rock. The porosity and permeability are directly influenced by the way the sand grains are packed together.

- Porosity is the percentage of bulk volume that is inhabited by interstices within a given rock. Porosity is directly influenced by the packing of even-sized spherical grains, rearranged from loosely packed to tightest packed in sandstones.

- Permeability is the rate in which water or other fluids flow through the rock. For groundwater, work permeability may be measured in gallons per day through a one square foot cross section under a unit hydraulic gradient.

## Types of Sandstone

All sandstones are composed of the same general minerals. These minerals make up the framework components of the sandstones. Such components are quartz, feldspars, and lithic fragments. Matrix may also be present in the interstitial spaces between the framework grains. These groups are divided based on mineralogy and texture. Even though sandstones have very simple compositions which are based on framework grains, geologists have not been able to agree on a specific, right way, to classify sandstones. Sandstone classifications are typically done by point-counting a thin section using a method like the Gazzi-Dickinson Method. The composition of a sandstone can have important information regarding the genesis of the sediment when used with a triangular *Q*uartz, *F*eldspar, *L*ithic fragment. Many geologists, however, do not agree on how to separate the triangle parts into the single components so that the framework grains can be plotted. Therefore, there have been many published ways to classify sandstones, all of which are similar in their general format.

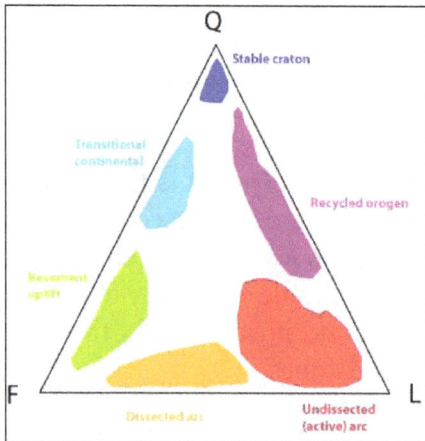

Schematic QFL diagram showing tectonic provinces.

Cross-bedding and scource in sandstone.

Visual aids are diagrams that allow geologists to interpret different characteristics about a sandstone. The following QFL chart and the sandstone provenance model correspond with each other therefore, when the QFL chart is plotted those points can then be plotted on the sandstone provenance model. The stage of textural maturity chart illustrates the different stages that a sandstone goes through.

- A QFL chart is a representation of the framework grains and matrix that is present in a sandstone. This chart is similar to those used in igneous petrology. When plotted correctly, this model of analysis creates for a meaningful quantitative classification of sandstones.

- A sandstone provenance chart allows geologists to visually interpret the different types of places from which sandstones can originate.

- A stage of textural maturity is a chart that shows the different stages of sandstones. This chart shows the difference between immature, submature, mature, and supermature sandstones. As the sandstone becomes more mature, grains become more rounded, and there is less clay in the matrix of the rock.

## Dott's Classification Scheme

Dott's sandstone classification scheme is one of many such schemes used by geologists for classifying sandstones. Dott's scheme is a modification of Gilbert's classification of silicate sandstones, and it incorporates R. L. Folk's dual textural and compositional maturity concepts into one classification system. The philosophy behind combining Gilbert's and R. L. Folk's schemes is that it is better able to "portray the continuous nature of textural variation from mudstone to arenite and from stable to unstable grain composition". Dott's classification scheme is based on the mineralogy of framework grains, and on the type of matrix present in between the framework grains.

In this specific classification scheme, Dott has set the boundary between arenite and wackes at 15% matrix. In addition, Dott also breaks up the different types of framework grains that can be present in a sandstone into three major categories: quartz, feldspar, and lithic grains.

- Arenites are types of sandstone that have less than 15% clay matrix in between the framework grains.

  ○ Quartz arenites are sandstones that contain more than 90% of siliceous grains. Grains can include quartz or chert rock fragments. Quartz arenites are texturally mature to supermature sandstones. These pure quartz sands result from extensive weathering that occurred before and during transport. This weathering removed everything but quartz grains, the most stable mineral. They are commonly affiliated with rocks that are deposited in a stable cratonic environment, such as aeolian beaches or shelf environments. Quartz arenites emanate from multiple recycling of quartz grains, generally as sedimentary source rocks and less regularly as first-cycle deposits derived from primary igneous or metamorphic rocks.

  ○ Feldspathic arenites are sandstones that contain less than 90% quartz, and more feldspar than unstable lithic fragments, and minor accessory minerals. Feldspathic sandstones are commonly immature or sub-mature. These sandstones occur in association with cratonic or stable shelf settings. Feldspathic sandstones are derived from granitic-type, primary crystalline, rocks. If the sandstone is dominantly plagioclase, then it is igneous in origin.

  ○ Lithic arenites are characterised by generally high content of unstable lithic fragments. Examples include volcanic and metamorphic clasts, though stable clasts such as chert are common in lithic arenites. This type of rock contains less than 90% quartz grains and more unstable rock fragments than feldspars. They are commonly immature to submature texturally. They are associated with fluvial conglomerates and other fluvial deposits, or in deeper water marine conglomerates. They are formed under conditions that produce large volumes of unstable material, derived from fine-grained rocks, mostly shales, volcanic rocks, and metamorphic rock.

- Wackes are sandstones that contain more than 15% clay matrix between framework grains.

  ○ Quartz wackes are uncommon because quartz arenites are texturally mature to supermature.

  ○ Felspathic wackes are feldspathic sandstone that contain a matrix that is greater than 15%.

  ○ Lithic wacke is a sandstone in which the matrix greater than 15%.

- Arkose sandstones are more than 25 percent feldspar. The grains tend to be poorly rounded and less well sorted than those of pure quartz sandstones. These feldspar-rich sandstones come from rapidly eroding granitic and metamorphic terrains where chemical weathering is subordinate to physical weathering.

- Greywacke sandstones are a heterogeneous mixture of lithic fragments and angular grains of quartz and feldspar or grains surrounded by a fine-grained clay matrix. Much of this matrix is formed by relatively soft fragments, such as shale and some volcanic rocks, that are chemically altered and physically compacted after deep burial of the sandstone formation.

## Uses

The Main Quadrangle of the University of Sydney, a so-called sandstone university.

17,000 yr old sandstone oil lamp discovered at the caves of Lascaux, France.

Sandstone statue Maria Immaculata by Fidelis Sporer, around 1770, in Freiburg, Germany.

Sandstone is highly absorbent. These are sandstone beverage coasters.

Weathered sandstone wall in Bayreuth, Germany.

Sandstone has been used for domestic construction and housewares since prehistoric times, and continues to be used.

Sandstone was a popular building material from ancient times. It is relatively soft, making it easy to carve. It has been widely used around the world in constructing temples, homes, and other buildings. It has also been used for artistic purposes to create ornamental fountains and statues.

Some sandstones are resistant to weathering, yet are easy to work. This makes sandstone a common building and paving material including in asphalt concrete. However, some that have been used in the past, such as the Collyhurst sandstone used in North West England, have been found less resistant, necessitating repair and replacement in older buildings. Because of the hardness of individual grains, uniformity of grain size and friability of their structure, some types of sandstone are excellent materials from which to make grindstones, for sharpening blades and other implements. Non-friable sandstone can be used to make grindstones for grinding grain, e.g., gritstone.

# MUDROCK

Mudrocks are a class of fine grained siliciclastic sedimentary rocks. The varying types of mudrocks include: *siltstone*, *claystone*, *mudstone*, *slate*, and *shale*. Most of the particles of which the stone is composed are less than 0.0625 mm (1/16th mm or 0.0025 inches) and are too small to study readily in the field. At first sight the rock types look quite similar; however, there are important differences in composition and nomenclature. There has been a great deal of disagreement involving the classification of mudrocks. There are a few important hurdles to classification, including:

1.  Mudrocks are the least understood, and one of the most understudied sedimentary rocks to date.

2.  It is difficult to study mudrock constituents, due to their diminutive size and susceptibility to weathering on outcrops.

3.  And most importantly, there is more than one classification scheme accepted by scientists.

Mudrocks make up fifty percent of the sedimentary rocks in the geologic record, and are easily the most widespread deposits on Earth. Fine sediment is the most abundant product of erosion, and these sediments contribute to the overall omnipresence of mudrocks. With increased pressure over time the platey clay minerals may become aligned, with the appearance of parallel layering (fissility). This finely bedded material that splits readily into thin layers is called *shale*, as distinct from *mudstone*. The lack of fissility or layering in mudstone may be due either to the original texture or to the disruption of layering by burrowing organisms in the sediment prior to lithification.

From the beginning of civilization, when pottery and mudbricks were made by hand, to now, mudrocks have been important. The first book on mudrocks, *Geologie des Argils* by Millot, was not published until 1964; however, scientists, engineers, and oil producers have understood the significance of mudrocks since the discovery of the Burgess Shale and the relatedness of mudrocks and oil. Literature on this omnipresent rock-type has been increasing in recent years, and technology continues to allow for better analysis.

Mudrocks, by definition, consist of at least fifty percent mud-sized particles. Specifically, mud is composed of silt-sized particles that are between $1/16 - 1/256$ $((1/16)^2)$ of a millimeter in diameter, and clay-sized particles which are less than 1/256 millimeter.

Mudrocks contain mostly clay minerals, and quartz and feldspars. They can also contain the following particles at less than 63 micrometres: calcite, dolomite, siderite, pyrite, marcasite, heavy minerals, and even organic carbon.

There are various synonyms for fine-grained siliciclastic rocks containing fifty percent or more of its constituents less than 1/256 of a millimeter. Mudstones, shales, lutites, and argillites are common qualifiers, or umbrella-terms; however, the term mudrock has increasingly become the terminology of choice by sedimentary geologists and authors.

The term "mudrock" allows for further subdivisions of siltstone, claystone, mudstone, and shale. For example, a siltstone would be made of more than 50-percent grains that equate to 1/16 - 1/256 of a millimeter. "Shale" denotes fissility, which implies an ability to part easily or break parallel to stratification. Siltstone, mudstone, and claystone implies lithified, or hardened, detritus without fissility.

Overall, "mudrocks" may be the most useful qualifying term, because it allows for rocks to be divided by its greatest portion of contributing grains and their respective grain size, whether silt, clay, or mud.

| Type | Min grain | Max grain |
|------|-----------|-----------|
| Claystone | 0 μm | 4 μm |
| Mudstone | 0 μm | 64 μm |
| Siltstone | 4 μm | 64 μm |
| Shale | 0 μm | 64 μm |
| Slate | na | na |

# Claystone

Claystone in Slovakia.

A claystone is lithified, and non-fissile mudrock. In order for a rock to be considered a claystone, it must consist of up to fifty percent clay, which measures less than 1/256 of a millimeter in particle size. Clay minerals are integral to mudrocks, and represent the first or second most abundant constituent by volume. There are 35 recognized clay mineral species on Earth. They make muds cohesive and plastic, or able to flow. Clay is by far the smallest particles recognized in mudrocks. Most materials in nature are clay minerals, but quartz, feldspar, iron oxides, and carbonates can weather to sizes of a typical clay mineral.

For a size comparison, a clay-sized particle is 1/1000 the size of a sand grain. This means a clay particle will travel 1000 times further at constant water velocity, thus requiring quieter conditions for settlement.

The formation of clay is well understood, and can come from soil, volcanic ash, and glaciation. Ancient mudrocks are another source, because they weather and disintegrate easily. Feldspar, amphiboles, pyroxenes, and volcanic glass are the principle donors of clay minerals.

# Mudstone

A mudstone is a siliciclastic sedimentary rock that contains a mixture of silt- and clay-sized particles (at least 1/3 of each).

The terminology of "mudstone" is not to be confused with the Dunham classification scheme for limestones. In Dunham's classification, a mudstone is any limestone containing less than ten percent carbonate grains. Note, a siliciclastic mudstone does not deal with carbonate grains. Friedman, Sanders, and Kopaska-Merkel suggest the use of "lime mudstone" to avoid confusion with siliciclastic rocks.

## Siltstone

Siltstone at UAT and Estonia.

A siltstone is a lithified, non-fissile mudrock. In order for a rock to be named a siltstone, it must contain over fifty percent silt-sized material. Silt is any particle smaller than sand, 1/16 of a millimeter, and larger than clay, 1/256 of millimeter. Silt is believed to be the product of physical weathering, which can involve freezing and thawing, thermal expansion, and release of pressure. Physical weathering does not involve any chemical changes in the rock, and it may be best summarised as the physical breaking apart of a rock.

One of the highest proportions of silt found on Earth is in the Himalayas, where phyllites are exposed to rainfall of up to five to ten meters (16 to 33 feet) a year. Quartz and feldspar are the biggest contributors to the silt realm, and silt tends to be non-cohesive, non-plastic, but can liquefy easily.

There is a simple test that can be done in the field to determine whether a rock is a siltstone or not, and that is to put the rock to one's teeth. If the rock feels "gritty" against one's teeth, then it is a siltstone.

## Shale

Marcellus Shale, New York.

Black Shale with pyrite.

Shale is a fine grained, hard, laminated mudrock, consisting of clay minerals, and quartz and feldspar silt. Shale is lithified and fissile. It must have at least 50-percent of its particles measure less than 0.062 mm. This term is confined to argillaceous, or clay-bearing, rock.

There are many varieties of shale, including calcareous and organic-rich; however, black shale, or organic-rich shale, deserves further evaluation. In order for a shale to be a black shale, it must contain more than one percent organic carbon. A good source rock for hydrocarbons can contain up to twenty percent organic carbon. Generally, black shale receives its influx of carbon from algae, which decays and forms an ooze known as sapropel. When this ooze is cooked at desired pressure, three to six kilometers (1.8 - 3.7 miles) depth, and temperature, 90–120 °C (194–248 °F), it will form kerogen. Kerogen can be heated, and yield up to 10–150 US gallons (0.038–0.568 m³) of natural oil & gas product per ton of rock.

## Slate

Slate Roof.

Slate is a hard mudstone that has undergone metamorphism, and has well-developed cleavage. It has gone through metamorphism at temperatures between 200–250 °C (392–482 °F), or extreme deformation. Since slate is formed in the lower realm of metamorphism, based on pressure and temperature, slate retains its stratification and can be defined as a hard, fine-grained rock.

Slate is often used for roofing, flooring, or old-fashioned stone walls. It has an attractive appearance, and its ideal cleavage and smooth texture are desirable.

## Creation of Mud and Mudrocks

Most mudrocks form in oceans or lakes, because these environments provide the quiet waters necessary for deposition. Although mudrocks can be found in every depositional environment on Earth, the majority are found in lakes and oceans.

## Mud Transport and Supply

Heavy rainfall provides the kinetic motion necessary for mud, clay, and silt transport. Southeast Asia, including Bangladesh and India, receives high amounts of rain from monsoons, which then washes sediment from the Himalayas and surrounding areas to the Indian Ocean.

Warm, wet climates are best for weathering rocks, and there is more mud on ocean shelves off tropical coasts than on temperate or polar shelves. The Amazon system, for example, has the third largest sediment load on Earth, with rainfall providing clay, silt, and mud from the Andes in Peru, Ecuador, and Bolivia.

Rivers, waves, and longshore currents segregate mud, silt, and clay from sand and gravel due to fall velocity. Longer rivers, with low gradients and large watersheds, have the best carrying capacity for mud. The Mississippi River, a good example of long, low gradient river with a large amount of water, will carry mud from its northernmost sections, and deposit the material in its mud-dominated delta.

## Mudrock Depositional Environments

## Alluvial Environments

The Ganges in India, the Yellow in China, and the Lower Mississippi in the United States are good examples of alluvial valleys. These systems have a continuous source of water, and can contribute mud through overbank sedimentation, when mud and silt is deposited overbank during flooding, and oxbow sedimentation where an abandoned stream is filled by mud.

In order for an alluvial valley to exist there must be a highly elevated zone, usually uplifted by active tectonic movement, and a lower zone, which acts as a conduit for water and sediment to the ocean.

## Glaciers

Vast quantities of mud and till are generated by glaciations and deposited on land as till and in lakes. Glaciers can erode already susceptible mudrock formations, and this process enhances glacial production of clay and silt.

The Northern Hemisphere contains 90-percent of the world's lakes larger than 500 km (310 mi), and glaciers created many of those lakes. Lake deposits formed by glaciation, including deep glacial scouring, are abundant.

## Non-Glacial Lakes

Although glaciers formed 90-percent of lakes in the Northern Hemisphere, they are not responsible for the formation of ancient lakes. Ancient lakes are the largest and deepest in the world, and hold up to twenty percent of today's petroleum reservoirs. They are also the second most abundant source of mudrocks, behind marine mudrocks.

Ancient lakes owe their abundance of mudrocks to their long lives and thick deposits. These deposits were susceptible to changes in oxygen and rainfall, and offer a robust account of paleoclimate consistency.

## Deltas

The Mississippi Delta.

A delta is a subaerial or subaqueous deposit formed where rivers or streams deposit sediment into a water body. Deltas, such as the Mississippi and Congo, have massive potential for sediment deposit, and can move sediments into deep ocean waters. Delta environments are found at the mouth of a river, where its waters slow as they enter the ocean, and silt and clay are deposited.

Low energy deltas, which deposit a great deal of mud, are located in lakes, gulfs, seas, and small oceans, where coastal currents are also low. Sand and gravel-rich deltas are high-energy deltas, where waves dominate, and mud and silt are carried much farther from the mouth of the river.

## Coastlines

Coastal currents, mud supply, and waves are a key factor in coastline mud deposition. The Amazon River supplies 500 million tons of sediment, which is mostly clay, to the coastal region of northeastern South America. 250 tons of this sediment moves along the coast and is deposited. Much of the mud accumulated here is more than 20 meters (65 feet) thick, and extends 30 kilometers (19 mi) into the ocean.

Much of the sediment carried by the Amazon can come from the Andes mountains, and the final distance traveled by the sediment is 6,000 km (3,700 mi).

## Marine Environments

70-percent of the Earth's surface is covered by ocean, and marine environments are where we find the world's highest proportion of mudrocks. There is a great deal of lateral continuity found in the ocean, as opposed to continents which are confined.

In comparison, continents are temporary stewards of mud and silt, and the inevitable home of mudrock sediments is the oceans. Reference the mudrock cycle below in order to understand the burial and resurgence of the various particles.

There are various environments in the oceans, including deep-sea trenches, abyssal plains, volcanic seamounts, convergent, divergent, and transform plate margins. Not only is land a major source of the ocean sediments, but organisms living within the ocean contribute, as well.

The world's rivers transport the largest volume of suspended and dissolved loads of clay and silt to the sea, where they are deposited on ocean shelves. At the poles, glaciers and floating ice drop deposits directly to the sea floor. Winds can provide fine grained material from arid regions, and explosive volcanic eruptions contribute as well. All of these sources vary in the rate of their contribution.

Sediment moves to the deeper parts of the oceans by gravity, and the processes in the ocean are comparable to those on land.

Location has a large impact on the types of mudrocks found in ocean environments. For example, the Apalachicola River, which drains in the subtropics of the United States, carries up to sixty to eighty percent kaolinite mud, whereas the Mississippi carries only ten to twenty percent kaolinite.

## The Mudrock Cycle

We can imagine the beginning of a mudrock's life as sediment at the top of a mountain, which may have been uplifted by plate tectonics or propelled into the air from a volcano. This sediment is exposed to rain, wind, and gravity which batters and breaks apart the rock by weathering. The products of weathering, including particles ranging from clay to silt, to pebbles and boulders, are transported to the basin below, where it can solidify into one if its many sedimentary mudstone types.

Eventually, the mudrock will move its way kilometers below the subsurface, where pressure and temperature cook the mudstone into a metamorphosed gneiss. The metamorphosed gneiss will make its way to the surface once again as country rock or as magma in a volcano, and the whole process will begin again.

## Important Properties

## Color

Mudrocks form in various colors, including: red, purple, brown, yellow, green and grey, and even black. Shades of grey are most common in mudrocks, and darker colors of black come from organic carbons. Green mudrocks form in reducing conditions, where organic matter decomposes along with ferric iron. They can also be found in marine environments, where pelagic, or free-floating species, settle out of the water and decompose in the mudrock. Red mudrocks form when iron within the mudrock becomes oxidized, and depending on the intensity of red, one can determine if the rock has fully oxidized.

## Fossils

Burgess Shale.

Fossils are well preserved in mudrock formations, because the fine-grained rock protects the fossils from erosion, dissolution, and other processes of erosion. Fossils are particularly important for recording past environments. Paleontologists can look at a specific area and determine salinity, water depth, water temperature, water turbidity, and sedimentation rates with the aid of type and abundance of fossils in mudrock.

One of the most famous mudrock formations is the Burgess Shale in Western Canada, which formed during the Cambrian. At this site, soft bodied creatures were preserved, some in whole, by the activity of mud in a sea. Solid skeletons are, generally, the only remnants of ancient life preserved; however, the Burgess Shale includes hard body parts such as bones, skeletons, teeth, and also soft body parts such as muscles, gills, and digestive systems. The Burgess Shale is one of the most significant fossil locations on Earth, preserving innumerable specimens of 500 million year old species, and its preservation is due to the protection of mudrock.

Another noteworthy formation is the Morrison Formation. This area covers 1.5 million square miles, stretching from Montana to New Mexico in the United States. It is considered one of the world's most significant dinosaur burial grounds, and its many fossils can be found in museums around the world. This site includes dinosaur fossils from a few dinosaur species, including the Allosaurus, Diplodocus, Stegosaurus, and Brontosaurus. There are also lungfish, freshwater mollusks, ferns and conifers. This deposit was formed by a humid, tropical climate with lakes, swamps, and rivers, which deposited mudrock. Inevitably, mudrock preserved countless specimens from the late Jurassic, roughly 150 million years ago.

## Petroleum and Natural Gas

Mudrocks, especially black shale, are the source and containers of precious petroleum sources throughout the world. Since mudrocks and organic material require quiet water conditions for deposition, mudrocks are the most likely resource for petroleum. Mudrocks have low porosity, they are impermeable, and often, if the mudrock is not black shale, it remains useful as a seal to petroleum and natural gas reservoirs. In the case of petroleum found in a reservoir, the rock surrounding the petroleum is not the source rock, whereas black shale is a source rock.

## Importance

As noted before, mudrocks make up fifty percent of the Earth's sedimentary geological record. They are widespread on Earth, and important for various industries.

Metamorphosed shale can hold emerald and gold, and mudrocks can host ore metals such as lead and zinc. Mudrocks are important in the preservation of petroleum and natural gas, due to their low porosity, and are commonly used by engineers to inhibit harmful fluid leakage from landfills.

Sandstones and carbonates record high-energy events in our history, and they are much easier to study. Interbedded between the high-energy events are mudrock formations that have recorded quieter, normal conditions in our Earth's history. It is the quieter, normal events of our geologic history we don't yet understand. Sandstones provide the big tectonic picture and some indications of water depth; mudrocks record oxygen content, a generally richer fossil abundance and diversity, and a much more informative geochemistry.

## Shale Rock

Shale is the most common sedimentary rock, accounting for about 70 percent of the rock found in the Earth's crust. It is a fine-grained clastic sedimentary rock made of compacted mud consisting of clay and tiny particles of quartz, calcite, mica, pyrite, other minerals, and organic compounds. Shale is found all over the world in areas where water exists or once flowed.

## How Shale Forms

Shale forms via compaction, typically from particles in slow or quiet water, such as river deltas, lakes, swamps, or the ocean floor. Heavier particles sink, ultimately forming sandstone and limestone, while clay and fine silt remain suspended in water. Over time, these fine particles settle and build upon each other, forming rock. Shale typically occurs in a broadsheet, several meters thick. Depending on the geography, lenticular formations may also form. Sometimes animal tracks, fossils, or even imprints of raindrops are preserved in shale layers.

## Composition and Properties

The clay clasts or particles in shale are less than 0.004 millimeters in diameter, which means the structure of the rock only becomes visible under magnification. The clay comes from decomposition of feldspar. Shale consists of at least 30 percent clay, with varying amounts of quartz, feldspar, carbonates, iron oxides, and organic matter. Oil shale or bituminous also contains kerogen, a mixture of hydrocarbons from deceased plants and animals. Shale tends to be classified based on its mineral content, so there is siliceous shale (silica), calcareous shale (calcite or dolomite), limonitic or hematitic shale (iron minerals), carbonaceous or bituminous shale (carbon compounds), and phospatic shale (phosphate).

The color of shale depends on the composition of the minerals. Shale with a higher organic (carbon) content tends to be darker in color and may be black or gray. The presence of ferric iron compounds yields red, brown, or purple shale. Ferrous iron yields black, blue, and green shale. Shale containing a lot of calcite tends to be pale gray or yellow.

The grain size and composition of minerals in shale determine its permeability, hardness, and plasticity. In general, shale is fissile and readily splits into layers parallel to the bedding plane, which is the plane of clay flake deposition. Shale is laminated, meaning the rock consists of many thin layers that are bound together.

## Commercial Uses

Shale has many commercial uses. It is a source material in the ceramics industry to make brick, tile, and pottery. Shale used to make pottery and building materials requires little processing besides crushing and mixing with water.

Shale is crushed and heated with limestone to make cement for the construction industry. Heating drives off water and breaks limestone into calcium oxide and carbon dioxide. Carbon dioxide is lost as a gas, leaving calcium oxide and clay, which hardens when mixed with water and allowed to dry.

The petroleum industry uses fracking to extract oil and natural gas from oil shale. Fracking involves injection of liquid at high pressure into the rock to force out the organic molecules. Typically high temperatures and special solvents are needed to extract the hydrocarbons, leading to waste products that raise concerns about environmental impact.

## Shale, Slate and Schist

Up to the mid-19th century, the term "slate" was often used to refer to shale, slate, *and*schist. Underground coal miners may still refer to shale as slate, per tradition. These sedimentary rocks have the same chemical composition and may occur together. The initial sedimentation of particles forms sandstone and mudstone. Shale forms when the mudstone becomes laminated and fissile. If shale is subjected to heat and pressure, it can metamorphose into slate. Slate can become phyllite, then schist, and eventually gneiss.

# PHOSPHORITE

Phosphorite, phosphate rock or rock phosphate is a non-detrital sedimentary rock which contains high amounts of phosphate minerals. The phosphate content of phosphorite (or grade of phosphate rock) varies greatly, from 4% to 20% phosphorus pentoxide ($P_2O_5$). Marketed phosphate rock is enriched ("beneficiated") to at least 28%, often more than 30% $P_2O_5$. This occurs through washing, screening, de-liming, magnetic separation or flotation. By comparison, the average phosphorus content of sedimentary rocks is less than 0.2%. The phosphate is present as fluorapatite $Ca_5(PO_4)_3F$ typically in cryptocrystalline masses (grain sizes < 1 μm) referred to as collophane-sedimentary apatite deposits of uncertain origin. It is also present as hydroxyapatite $Ca_5(PO_4)_3OH$ or $Ca_{10}(PO_4)_6(OH)_2$, which is often dissolved from vertebrate bones and teeth, whereas fluorapatite can originate from hydrothermal veins. Other sources also include chemically dissolved phosphate minerals from igneous and metamorphic rocks. Phosphorite deposits often occur in extensive layers, which cumulatively cover tens of thousands of square kilometres of the Earth's crust.

Limestones and mudstones are common phosphate-bearing rocks. Phosphate rich sedimentary rocks can occur in dark brown to black beds, ranging from centimeter-sized laminae to beds that are several meters thick. Although these thick beds can exist, they are rarely only composed of phosphatic sedimentary rocks. Phosphatic sedimentary rocks are commonly accompanied by or interbedded with shales, cherts, limestone, dolomites and sometimes sandstone. These layers contain the same textures and structures as fine-grained limestones and may represent diagenetic replacements of carbonate minerals by phosphates. They also can be composed of peloids, ooids, fossils, and clasts that are made up of apatite. There are some phosphorites that are very small and have no distinctive granular textures. This means that their textures are similar to that of collophane, or fine micrite-like texture. Phosphatic grains may be accompanied by organic matter, clay minerals, silt sized detrital grains, and pyrite. Peloidal or pelletal phosphorites occur normally; whereas oolitic phosphorites are not common.

Phosphorites are known from Proterozoic banded iron formations in Australia, but are more common from Paleozoic and Cenozoic sediments. The Permian Phosphoria Formation of the western United States represents some 15 million years of sedimentation. It reaches a thickness of 420 metres and covers an area of 350,000 km². Commercially mined phosphorites occur in France, Belgium, Spain, Morocco, Tunisia and Algeria. In the United States phosphorites have been mined in Florida, Tennessee, Wyoming, Utah, Idaho and Kansas.

## Classification of Phosphatic Sedimentary Rocks

- Pristine: Phosphates that are in pristine conditions have not undergone bioturbation. In other words, the word pristine is used when phosphatic sediment, phosphatized stromatolites and phosphate hardgrounds have not been disturbed.

- Condensed: Phosphatic particles, laminae and beds are considered condensed when they have been concentrated. This is helped by the extracting and reworking processes of phosphatic particles or bioturbation.

- Allochthonous: Phosphatic particles that were moved by turbulent or gravity-driven flows and deposited by these flows.

## Phosphorus Cycle, Formation and Accumulation

The heaviest accumulation of phosphorus is mainly on the ocean floor. Phosphorus accumulation occurs from atmospheric precipitation, dust, glacial runoff, cosmic activity, underground hydrothermal volcanic activity, and deposition of organic material. The primary inflow of dissolved phosphorus is from continental weathering, brought out by rivers to the ocean. It is then processed by both micro- and macro-organisms. Diatomaceous plankton, phytoplankton, and zooplankton process and dissolve phosphorus in the water. The bones and teeth of certain fish (e.g. anchovies) absorb phosphorus and are later deposited and buried in the ocean sediment.

Depending on the pH and salinity levels of the ocean water, organic matter will decay and releases phosphorus from sediment in shallow basins. Bacteria and enzymes dissolve organic matter on the water bottom interface, thus returning phosphorus to the beginning of its biogenic cycle. Mineralization of organic matter can also cause the release of phosphorus back into the ocean water.

## Depositional Environments

Phosphates are known to be deposited in a wide range of depositional environments. Normally phosphates are deposited in very shallow, near-shore marine or low energy environments. This includes environments such as supratidal zones, littoral or intertidal zones, and most importantly estuarine. Currently, areas of oceanic upwelling cause the formation of phosphates. This is because of the constant stream of phosphate brought from the large, deep ocean reservoir. This cycle allows continuous growth of organisms.

- Supratidal zones: Supratidal environments are part of the tidal flat system where the presence of strong wave activity is non-existent. Tidal flat systems are created along open coasts and relatively low wave energy environments. They can also develop on high energy coasts behind barrier islands where they are sheltered from the high energy wave action. Within the tidal flat system, the supratidal zone lies in a very high tide level. However, it can be flooded by extreme tides and cut across by tidal channels. This is also subaerially exposed, but is flooded twice a month by spring tides.

- Littoral environments/ intertidal zones: Intertidal zones are also part of the tidal flat system. The intertidal zone is located within the mean high and low tide levels. It is subject to tidal shifts, which means that it is subaerially exposed once or twice a day. It is not exposed long enough to withhold vegetation. The zone contains both suspension sedimentation and bed load.

- Estuarine environments: Estuarine environments, or estuaries, are located at the lower parts of rivers that flow into the open sea. Since they are in the seaward section of the drowned valley system they receive sediment from both marine and fluvial sources. These contain facies that are affected by tide and wave fluvial processes. An estuary is considered to stretch from the landward limit of tidal facies to the seaward limit of coastal facies. Phosphorites are often deposited in fjords within estuarine environments. These are estuaries with shallow sill constrictions. During Holocene sea-level rise, fjord estuaries formed by drowning of glacially-eroded U-shaped valleys.

The most common occurrence of phosphorites is related to strong marine upwelling of sediments. Upwelling is caused by deep water currents that are brought up to coastal surfaces where a large deposition of phosphorites may occur. This type of environment is the main reason why phosphorites are commonly associated with silica and chert. Estuaries are also known as a phosphorus "trap". This is because coastal estuaries contain a high productivity of phosphorus from marsh grass and benthic algae which allow an equilibrium exchange between living and dead organisms.

## Types of Phosphorite Deposition

- Phosphate nodules: These are spherical concentrations that are randomly distributed along the floor of continental shelves. Most phosphorite grains are sand size although particles greater than 2 mm may be present. These larger grains, referred to as nodules, can range up to several tens of centimeters in size.

- Bioclastic phosphates or bone beds: Bone beds are bedded phosphate deposits that contain concentrations of small skeletal particles and coprolites. Some also contain invertebrate fossils like brachiopods and become more enriched in $P_2O_5$ after diagentic processes have occurred. Bioclastic phosphates can also be cemented by phosphate minerals.

- Phosphatization: Phosphatization is a type of rare diagenetic processes. It occurs when fluids that are rich in phosphate are leached from guano. These are then concentrated and reprecipitated in limestone. Phosphatized fossils or fragments of original phosphatic shells are important components within some these deposits.

## Tectonic and Oceanographic Settings of Marine Phosphorites

- Epeiric sea phosphorites: Epeiric sea phosphorites are within marine shelf environments. These are in a broad and shallow cratonic setting. This is where granular phosphorites, phosphorite hardgrounds, and nodules occur.

- Continental margin phosphorites: Convergent, passive, upwelling, non-upwelling. This environment accumulates phosphorites in the form of hardgrounds, nodules and granular beds. These accumulate by carbonate fluorapatite percipitaion during early diagenesis in the upper few tens of centimeters of sediment. There are two different environmental conditions in which phosphorites are produced within continental margins. Continental margins can consist of organic rich sedimentation, strong coastal upwelling, and pronounced low oxygen zones. They can also form in conditions such as oxygen rich bottom waters and organic poor sediments.

- Seamount phosphorites: These are phosphorites that occur in seamounts, guyots, or flat topped seamounts, seamount ridges. These phosphorites are produced in association with iron and magnesium bearing crusts. In this setting the productivity of phosphorus is recycled within an iron oxidation reduction phosphorus cycle. This cycle can also form glauconite which is normally associated with modern and ancient phosphorites.

- Insular phosphorites: Insular phosphorites are located in carbonate islands, plateaus, coral island consisting of a reef surrounding a lagoon or, atoll lagoon, marine lakes. The phosphorite here originates from guano. Replacement of deep sea sediments precipitates, that has been formed in place on the ocean floor.

## Production and Use

Guano phosphorite mining in the Chincha Islands of Peru, c. 1860.

Phosphorite mine near Oron, Negev, Israel.

Deposits which contain phosphate in quantity and concentration which are economic to mine as ore for their phosphate content are not particularly common. The two main sources for phosphate are guano, formed from bird droppings, and rocks containing concentrations of the calcium phosphate mineral, apatite.

Phosphate rock is mined, beneficiated, and either solubilized to produce wet-process phosphoric acid, or smelted to produce elemental phosphorus. Phosphoric acid is reacted with phosphate rock to produce the fertilizer triple superphosphate or with anhydrous ammonia to produce the ammonium phosphate fertilizers. Elemental phosphorus is the base for furnace-grade phosphoric acid, phosphorus pentasulfide, phosphorus pentoxide, and phosphorus trichloride. Approximately 90% of phosphate rock production is used for fertilizer and animal feed supplements and the balance for industrial chemicals.

Froth flotation is used to concentrate the mined phosphorus to rock phosphate. The mined ore slurry is treated with fatty acids to cause calcium phosphate to become hydrophobic.

For general use in the fertilizer industry, phosphate rock or its concentrates preferably have levels of 30% phosphorus pentoxide ($P_2O_5$), reasonable amounts of calcium carbonate (5%), and <4% combined iron and aluminium oxides. Worldwide, the resources of high-grade ore are declining, and the beneficiation of lower grade ores by washing, flotation and calcining is becoming more widespread.

In addition to phosphate fertilisers for agriculture, phosphorus from rock phosphate is also used in animal feed supplements, food preservatives, anti-corrosion agents, cosmetics, fungicides, ceramics, water treatment and metallurgy.

# LIMESTONE

Limestone is a carbonate sedimentary rock that is often composed of the skeletal fragments of marine organisms such as coral, foraminifera, and molluscs. Its major materials are the minerals calcite and aragonite, which are different crystal forms of calcium carbonate ($CaCO_3$). A closely related rock is dolomite, which contains a high percentage of the mineral dolomite, $CaMg(CO_3)_2$.

In old USGS publications, dolomite was referred to as magnesian limestone, a term now reserved for magnesium-deficient dolomites or magnesium-rich limestones.

Travertine limestone terraces of Pamukkale, Turkey.

About 10% of sedimentary rocks are limestones. The solubility of limestone in water and weak acid solutions leads to karst landscapes, in which water erodes the limestone over thousands to millions of years. Most cave systems are through limestone bedrock.

Limestone has numerous uses: as a building material, an essential component of concrete (Portland cement), as aggregate for the base of roads, as white pigment or filler in products such as toothpaste or paints, as a chemical feedstock for the production of lime, as a soil conditioner, or as a popular decorative addition to rock gardens.

Limestone quarry at Cedar Creek, Virginia, USA.

Cutting limestone blocks at a quarry in Gozo, Malta.

Limestone as building material.

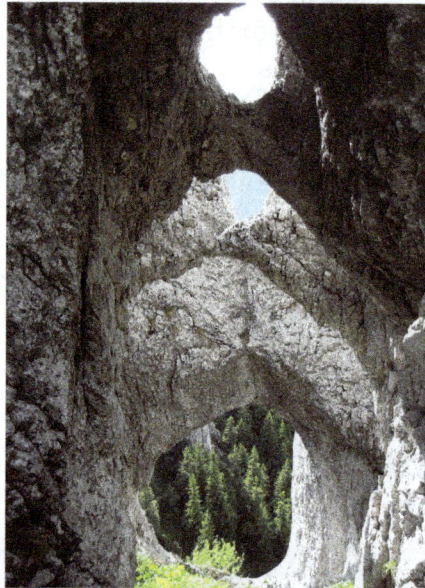

La Zaplaz formations in the Piatra Craiului Mountains, Romania.

Like most other sedimentary rocks, most limestone is composed of grains. Most grains in limestone are skeletal fragments of marine organisms such as coral or foraminifera. These organisms secrete shells made of aragonite or calcite, and leave these shells behind when they die. Other carbonate grains composing limestones are ooids, peloids, intraclasts, and extraclasts.

Limestone often contains variable amounts of silica in the form of chert (chalcedony, flint, jasper, etc.) or siliceous skeletal fragment (sponge spicules, diatoms, radiolarians), and varying amountsl precipitation of calcite or aragonite, i.e. travertine. Secondary calcite may be deposited by supersaturated meteoric waters (groundwater that precipitates the material in caves). This produces speleothems, such as stalagmites and stalactites. Another form taken by calcite is oolitic limestone, which can be recognized by its granular (oolite) appearance.

The primary source of the calcite in limestone is most commonly marine organisms. Some of these organisms can construct mounds of rock known as reefs, building upon past generations. Below about 3,000 meters, water pressure and temperature conditions cause the dissolution of calcite to increase nonlinearly, so limestone typically does not form in deeper waters. Limestones may also form in lacustrine and evaporite depositional environments.

Calcite can be dissolved or precipitated by groundwater, depending on several factors, including the water temperature, pH, and dissolved ion concentrations. Calcite exhibits an unusual characteristic called retrograde solubility, in which it becomes less soluble in water as the temperature increases.

Impurities (such as clay, sand, organic remains, iron oxide, and other materials) will cause limestones to exhibit different colors, especially with weathered surfaces.

Limestone may be crystalline, clastic, granular, or massive, depending on the method of formation. Crystals of calcite, quartz, dolomite or barite may line small cavities in the rock. When conditions are right for precipitation, calcite forms mineral coatings that cement the existing rock grains together, or it can fill fractures.

Travertine is a banded, compact variety of limestone formed along streams, particularly where there are waterfalls and around hot or cold springs. Calcium carbonate is deposited where evaporation of the water leaves a solution supersaturated with the chemical constituents of calcite. Tufa, a porous or cellular variety of travertine, is found near waterfalls. Coquina is a poorly consolidated limestone composed of pieces of coral or shells.

During regional metamorphism that occurs during the mountain building process (orogeny), limestone recrystallizes into marble. Limestone is a parent material of Mollisol soil group.

## Classification

Two major classification schemes, the Folk and the Dunham, are used for identifying the types of carbonate rocks collectively known as limestone.

Ooids from a beach on Joulter's Cay, The Bahamas.

Ooids in limestone of the Carmel Formation (Middle Jurassic) of southwestern Utah.

## Folk Classification

Robert L. Folk developed a classification system that places primary emphasis on the detailed composition of grains and interstitial material in carbonate rocks. Based on composition, there are three main components: allochems (grains), matrix (mostly micrite), and cement (sparite). The Folk system uses two-part names; the first refers to the grains and the second is the root. It is helpful to have a petrographic microscope when using the Folk scheme, because it is easier to determine the components present in each sample.

## Dunham Classification

The Dunham scheme focuses on depositional textures. Each name is based upon the texture of the grains that make up the limestone. Robert J. Dunham published his system for limestone in 1962; it focuses on the depositional fabric of carbonate rocks. Dunham divides the rocks into four main groups based on relative proportions of coarser clastic particles. Dunham names are essentially for rock families. His efforts deal with the question of whether or not the grains were originally in mutual contact, and therefore self-supporting, or whether the rock is characterized by the presence of frame builders and algal mats. Unlike the Folk scheme, Dunham deals with the original porosity of the rock. The Dunham scheme is more useful for hand samples because it is based on texture, not the grains in the sample.

A revised classification was proposed by Wright (1992). It adds some diagenetic patterns and can be summarized as follows:

| Revised Dunham classification (Wright 1992) | | | | | | | | | | |
|---|---|---|---|---|---|---|---|---|---|---|
| Depositional | | | | Biological | | | Diagenetic | | | |
| Matrix-supported (clay and silt grade) | | Grain-supported | | In situ organisms | | | Non-obliterative | | | Obliterative |
| Less than 10% grains | More than 10% grains | With matrix | No Matrix | Encrusting binding organisms | Organisms acted to baffle | Rigid organisms dominant | Main component is cement | Many grain contact as microstylolithes | Most grain contacts are microstylolithes | Crystals larger 10 micrometers |

| Carbonate mud-stone | Wacke-stone | Pack-stone | Grain-stone | Bound-stone | Baffle-stone | Frame-stone | Ce-ment-stone | Con-densed grainstone | Fitted grainstone | Sparstone |
|---|---|---|---|---|---|---|---|---|---|---|
| | Components larger 2 mm | | | Microsparstone | | | | | | Crystals smaller 10 micrometers |
| | Float-stone | Rudstone | | | | | | | | |

## Limestone Landscape

The Cudgel of Hercules, a tall limestone rock (Pieskowa Skała Castle in the background).

The Samulá cenote in Valladolid, Yucatán, Mexico.

Reflecting lake in the Luray Caverns of the northern Shenandoah Valley.

The White Cliffs of Dover.

About 10% of all sedimentary rocks are limestones. Limestone is partially soluble, especially in acid, and therefore forms many erosional landforms. These include limestone pavements, pot holes, cenotes, caves and gorges. Such erosion landscapes are known as karsts. Limestone is less resistant than most igneous rocks, but more resistant than most other sedimentary rocks. It is therefore usually associated with hills and downland, and occurs in regions with other sedimentary rocks, typically clays.

Karst topography and caves develop in limestone rocks due to their solubility in dilute acidic groundwater. The solubility of limestone in water and weak acid solutions leads to karst landscapes. Regions overlying limestone bedrock tend to have fewer visible above-ground sources (ponds and streams), as surface water easily drains downward through joints in the limestone. While draining, water and organic acid from the soil slowly (over thousands or millions of years) enlarges these cracks, dissolving the calcium carbonate and carrying it away in solution. Most cave systems are through limestone bedrock. Cooling groundwater or mixing of different groundwaters will also create conditions suitable for cave formation.

Coastal limestones are often eroded by organisms which bore into the rock by various means. This process is known as bioerosion. It is most common in the tropics, and it is known throughout the fossil record.

Bands of limestone emerge from the Earth's surface in often spectacular rocky outcrops and islands. Examples include the Rock of Gibraltar, the Burren in County Clare, Ireland; the Verdon Gorge in France; Malham Cove in North Yorkshire and the Isle of Wight, England; the Great Orme in Wales; on Fårö near the Swedish island of Gotland, the Niagara Escarpment in Canada/United States, Notch Peak in Utah, the Ha Long Bay National Park in Vietnam and the hills around the Lijiang River and Guilin city in China.

The Florida Keys, islands off the south coast of Florida, are composed mainly of oolitic limestone (the Lower Keys) and the carbonate skeletons of coral reefs (the Upper Keys), which thrived in the area during interglacial periods when sea level was higher than at present.

Unique habitats are found on alvars, extremely level expanses of limestone with thin soil mantles. The largest such expanse in Europe is the Stora Alvaret on the island of Öland, Sweden. Another area with large quantities of limestone is the island of Gotland, Sweden. Huge quarries in northwestern Europe, such as those of Mount Saint Peter (Belgium/Netherlands), extend for more than a hundred kilometers.

The world's largest limestone quarry is at Michigan Limestone and Chemical Company in Rogers City, Michigan.

## Uses

The Megalithic Temples of Malta such as Ħaġar Qim are built entirely of limestone.
They are among the oldest freestanding structures in existence.

Limestone is very common in architecture, especially in Europe and North America. Many landmarks across the world, including the Great Pyramid and its associated complex in Giza, Egypt, were made of limestone. So many buildings in Kingston, Ontario, Canada were, and continue to be, constructed from it that it is nicknamed the 'Limestone City'. On the island of Malta, a variety of limestone called Globigerina limestone was, for a long time, the only building material available, and is still very frequently used on all types of buildings and sculptures. Limestone is readily available and relatively easy to cut into blocks or more elaborate carving. Ancient American sculptors valued limestone because it was easy to work and good for fine detail. Going back to the Late Preclassic period (by 200–100 BCE), the Maya civilization (Ancient Mexico) created refined sculpture using limestone because of these excellent carving properties. The Maya would decorate the ceilings of their sacred buildings (known as lintels) and cover the walls with carved limestone panels. Carved on these sculptures were political and social stories, and this helped communicate messages of the king to his people. Limestone is long-lasting and stands up well to exposure, which explains why many limestone ruins survive. However, it is very heavy, making it impractical for tall buildings, and relatively expensive as a building material.

The Great Pyramid of Giza, one of the Seven Wonders of the Ancient World had an outside cover made entirely from limestone.

A limestone plate with a negative map of Moosburg in Bavaria is prepared for a lithography print.

Riley County Courthouse built of limestone in Manhattan, Kansas, USA.

Limestone was most popular in the late 19th and early 20th centuries. Train stations, banks and other structures from that era are normally made of limestone. It is used as a facade on some skyscrapers, but only in thin plates for covering, rather than solid blocks. In the United States, Indiana, most notably the Bloomington area, has long been a source of high quality quarried limestone, called Indiana limestone. Many famous buildings in London are built from Portland limestone.

Limestone was also a very popular building block in the Middle Ages in the areas where it occurred, since it is hard, durable, and commonly occurs in easily accessible surface exposures. Many medieval churches and castles in Europe are made of limestone. Beer stone was a popular kind of limestone for medieval buildings in southern England.

Limestone and (to a lesser extent) marble are reactive to acid solutions, making acid rain a significant problem to the preservation of artifacts made from this stone. Many limestone statues and building surfaces have suffered severe damage due to acid rain. Likewise limestone gravel has been used to protect lakes vulnerable to acid rain, acting as a pH buffering agent. Acid-based cleaning chemicals can also etch limestone, which should only be cleaned with a neutral or mild alkali-based cleaner.

Other uses include:

- It is the raw material for the manufacture of quicklime (calcium oxide), slaked lime (calcium hydroxide), cement and mortar.

- Pulverized limestone is used as a soil conditioner to neutralize acidic soils (agricultural lime).

- Is crushed for use as aggregate—the solid base for many roads as well as in asphalt concrete.

- Geological formations of limestone are among the best petroleum reservoirs.

- As a reagent in flue-gas desulfurization, it reacts with sulfur dioxide for air pollution control.

- Glass making, in some circumstances, uses limestone.

- It is added to toothpaste, paper, plastics, paint, tiles, and other materials as both white pigment and a cheap filler.

- It can suppress methane explosions in underground coal mines.

- Purified, it is added to bread and cereals as a source of calcium.

- Calcium levels in livestock feed are supplemented with it, such as for poultry (when ground up).

- It can be used for remineralizing and increasing the alkalinity of purified water to prevent pipe corrosion and to restore essential nutrient levels.

- Used in blast furnaces, limestone binds with silica and other impurities to remove them from the iron.

- It is used in sculptures because of its suitability for carving.

# EVAPORITE

Evaporite is the term for a water-soluble mineral sediment that results from concentration and crystallization by evaporation from an aqueous solution. There are two types of evaporite deposits: marine, which can also be described as ocean deposits, and non-marine, which are found in standing bodies of water such as lakes. Evaporites are considered sedimentary rocks and are formed by chemical sediments.

## Formation of Evaporite Rocks

Although all water bodies on the surface and in aquifers contain dissolved salts, the water must evaporate into the atmosphere for the minerals to precipitate. For this to happen, the water body must enter a restricted environment where water input into this environment remains below the net rate of evaporation. This is usually an arid environment with a small basin fed by a limited input of water. When evaporation occurs, the remaining water is enriched in salts, and they precipitate when the water becomes supersaturated.

## Evaporite Depositional Environments

### Marine Evaporites

Anhydrite.

Marine evaporites tend to have thicker deposits and are usually the focus of more extensive research. They also have a system of evaporation. When scientists evaporate ocean water in a laboratory, the minerals are deposited in a defined order that was first demonstrated by Usiglio in 1884. The first phase of the experiment begins when about 50% of the original water depth remains. At this point, minor carbonates begin to form. The next phase in the sequence comes when the experiment is left with about 20% of its original level. At this point, the mineral gypsum begins to form, which is then followed by halite at 10%, excluding carbonate minerals that tend not to be evaporites. The most common minerals that are generally considered to be the most representative of marine evaporites are calcite, gypsum and anhydrite, halite, sylvite, carnallite, langbeinite, polyhalite, and kainite. Kieserite ($MgSO_4$) may also be included, which often will make up less than four percent of the overall content. However, there are approximately 80 different minerals that have been reported found in evaporite deposits, though only about a dozen are common enough to be considered important rock formers.

## Non-Marine Evaporites

Non-marine evaporites are usually composed of minerals that are not common in marine environments because in general the water from which non-marine evaporite precipitates has proportions of chemical elements different from those found in the marine environments. Common minerals that are found in these deposits include blödite, borax, epsomite, gaylussite, glauberite, mirabilite, thenardite and trona. Non-marine deposits may also contain halite, gypsum, and anhydrite, and may in some cases even be dominated by these minerals, although they did not come from ocean deposits. This, however, does not make non-marine deposits any less important; these deposits often help to paint a picture into past Earth climates. Some particular deposits even show important tectonic and climatic changes. These deposits also may contain important minerals that help in today's economy. Thick non-marine deposits that accumulate tend to form where evaporation rates will exceed the inflow rate, and where there is sufficient soluble supplies. The inflow also has to occur in a closed basin, or one with restricted outflow, so that the sediment has time to pool and form in a lake or other standing body of water. Primary examples of this are called "saline lake deposits". Saline lakes includes things such as perennial lakes, which are lakes that are there year-round, playa lakes, which are lakes that appear only during certain seasons, or any other terms that are used to define places that hold standing bodies of water intermittently or year-round. Examples of modern non-marine depositional environments include the Great Salt Lake in Utah and the Dead Sea, which lies between Jordan and Israel.

Evaporite depositional environments that meet the above conditions include:

- Graben areas and half-grabens within continental rift environments fed by limited riverine drainage, usually in subtropical or tropical environments.

    ◦ Example environments at the present that match this is the Denakil Depression, Ethiopia; Death Valley, California.

- Graben environments in oceanic rift environments fed by limited oceanic input, leading to eventual isolation and evaporation.

    ◦ Examples include the Red Sea, and the Dead Sea in Jordan and Israel.

- Internal drainage basins in arid to semi-arid temperate to tropical environments fed by ephemeral drainage.

    ◦ Example environments at the present include the Simpson Desert, Western Australia, the Great Salt Lake in Utah.

- Non-basin areas fed exclusively by groundwater seepage from artesian waters.

    ◦ Example environments include the seep-mounds of the Victoria Desert, fed by the Great Artesian Basin, Australia.

- Restricted coastal plains in regressive sea environments.

  ○ Examples include the sabkha deposits of Iran, Saudi Arabia, and the Red Sea; the Garabogazköl of the Caspian Sea.

- Drainage basins feeding into extremely arid environments.

  ○ Examples include the Chilean deserts, certain parts of the Sahara, and the Namib.

The most significant known evaporite depositions happened during the Messinian salinity crisis in the basin of the Mediterranean.

## Evaporitic Formations

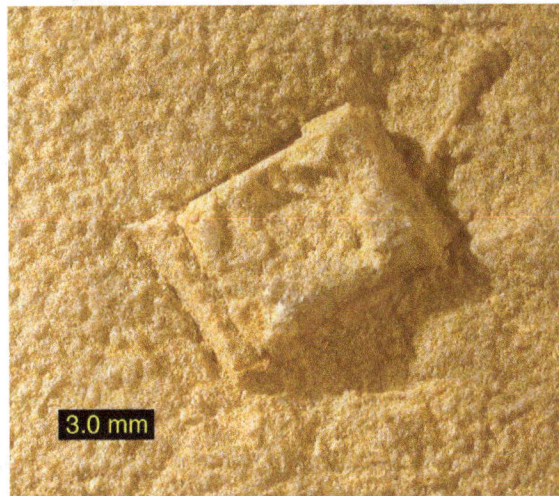

Hopper crystal cast of halite in a Jurassic rock, Carmel Formation, southwestern Utah.

Evaporite formations need not be composed entirely of halite salt. In fact, most evaporite formations do not contain more than a few percent of evaporite minerals, the remainder being composed of the more typical detrital clastic rocks and carbonates. Examples of evaporite formations include occurrences of evaporite sulfur in Eastern Europe and West Asia.

For a formation to be recognised as evaporitic it may simply require recognition of halite pseudomorphs, sequences composed of some proportion of evaporite minerals, and recognition of mud crack textures or other textures.

## Economic Importance of Evaporites

Evaporites are important economically because of their mineralogy, their physical properties in-situ, and their behaviour within the subsurface.

Evaporite minerals, especially nitrate minerals, are economically important in Peru and Chile. Nitrate minerals are often mined for use in the production on fertilizer and explosives.

Thick halite deposits are expected to become an important location for the disposal of nuclear waste because of their geologic stability, predictable engineering and physical behaviour, and imperviousness to groundwater.

Halite formations are famous for their ability to form diapirs, which produce ideal locations for trapping petroleum deposits.

Halite deposits are often mined for use as salt.

## Major Groups of Evaporite Minerals

Calcite.

This is a chart that shows minerals that form the marine evaporite rocks, they are usually the most common minerals that appear in this kind of deposit.

| Mineral class | Mineral name | Chemical Composition |
|---|---|---|
| Chlorides | Halite | $NaCl$ |
| | Sylvite | $KCl$ |
| | Carnallite | $KMgCl_3.6H_2O$ |
| | Kainite | $KMg(SO_4)Cl.3H_2O$ |
| Sulfates | Anhydrite | $CaSO_4$ |
| | Gypsum | $CaSO_4.2H_2O$ |
| | Kieserite | $MgSO_4.H_2O$ |
| | Langbeinite | $K_2Mg_2(SO_4)_3$ |
| | Polyhalite | $K_2Ca_2Mg(SO_4)_6.H_2O$ |
| Carbonates | Dolomite | $CaMg(CO_3)_2$ |
| | Calcite | $CaCO_3$ |
| | Magnesite | $MgCO_3$ |

Hanksite, $Na_{22}K(SO_4)_9(CO_3)_2Cl$, one of the few minerals that is both a carbonate and a sulfate

- Halides: Halite, sylvite (KCl), and fluorite

- Sulfates: Such as gypsum, barite, and anhydrite

- Nitrates: Nitratine (soda niter) and niter

- Borates: Typically found in arid-salt-lake deposits plentiful in the southwestern US. A common borate is borax, which has been used in soaps as a surfactant.

- Carbonates: Such as trona, formed in inland brine lakes.

  ○ Some evaporite minerals, such as Hanksite, are from multiple groups.

Evaporite minerals start to precipitate when their concentration in water reaches such a level that they can no longer exist as solutes.

The minerals precipitate out of solution in the reverse order of their solubilities, such that the order of precipitation from sea water is:

- Calcite ($CaCO_3$) and dolomite ($CaMg(CO_3)_2$)

- Gypsum ($CaSO_4 \cdot 2H_2O$) and anhydrite ($CaSO_4$).

- Halite (i.e. common salt, NaCl)

- Potassium and magnesium salts

The abundance of rocks formed by seawater precipitation is in the same order as the precipitation given above. Thus, limestone (calcite) and dolomite are more common than gypsum, which is more common than halite, which is more common than potassium and magnesium salts.

Evaporites can also be easily recrystallized in laboratories in order to investigate the conditions and characteristics of their formation.

## Salt

Halite commonly known as rock salt, is a type of salt, the mineral (natural) form of sodium chloride (NaCl). Halite forms isometric crystals. The mineral is typically colorless or white, but may also be light blue, dark blue, purple, pink, red, orange, yellow or gray depending on inclusion of other materials, impurities, and structural or isotopic abnormalities in the crystals. It commonly occurs with other evaporite deposit minerals such as several of the sulfates, halides, and borates.

## Occurrence

Halite cubes from the Stassfurt Potash deposit, Saxony-Anhalt, Germany.

Halite occurs in vast beds of sedimentary evaporite minerals that result from the drying up of enclosed lakes, playas, and seas. Salt beds may be hundreds of meters thick and underlie broad areas. In the United States and Canada extensive underground beds extend from the Appalachian basin of western New York through parts of Ontario and under much of the Michigan Basin. Other deposits are in Ohio, Kansas, New Mexico, Nova Scotia and Saskatchewan. The Khewra salt mine is a massive deposit of halite near Islamabad, Pakistan. Salt domes are vertical diapirs or pipe-like masses of salt that have been essentially "squeezed up" from underlying salt beds by mobilization due to the weight of overlying rock. Salt domes contain anhydrite, gypsum, and native sulfur, in addition to halite and sylvite. They are common along the Gulf coasts of Texas and Louisiana and are often associated with petroleum deposits. Germany, Spain, the Netherlands, Romania and Iran also have salt domes. Salt glaciers exist in arid Iran where the salt has broken through the surface at high elevation and flows downhill. In all of these cases, halite is said to be behaving in the manner of a rheid.

Unusual, purple, fibrous vein filling halite is found in France and a few other localities. Halite crystals termed hopper crystals appear to be "skeletons" of the typical cubes, with the edges present and stairstep depressions on, or rather in, each crystal face. In a rapidly crystallizing environment, the edges of the cubes simply grow faster than the centers. Halite crystals form very quickly in some rapidly evaporating lakes resulting in modern artifacts with a coating or encrustation of halite crystals. Halite flowers are rare stalactites of curling fibers of halite that are found in certain arid caves of Australia's Nullarbor Plain. Halite stalactites and encrustations are also reported in the Quincy native copper mine of Hancock, Michigan.

## Mining

The world's largest underground salt mine is the Sifto Salt Mine. It produces over 7 million tons of rock salt per year using the room and pillar mining method. It is located half a kilometre under Lake Huron in Ontario, Canada. In the United Kingdom there are three mines; the largest of these is at Winsford in Cheshire, producing, on average, one million tonnes of salt per year.

## Uses

Salt is used extensively in cooking as a flavor enhancer, and to cure a wide variety of foods such as bacon and fish. It is frequently used in food preservation methods across various cultures. Larger pieces can be ground in a salt mill or dusted over food from a shaker as finishing salt.

Halite is also often used both residentially and municipally for managing ice. Because brine (a solution of water and salt) has a lower freezing point than pure water, putting salt or saltwater on ice that is below 0 °C (32 °F) will cause it to melt. (This effect is called freezing-point depression.) It is common for homeowners in cold climates to spread salt on their sidewalks and driveways after a snow storm to melt the ice. It is not necessary to use so much salt that the ice is completely melted; rather, a small amount of salt will weaken the ice so that it can be easily removed by other means. Also, many cities will spread a mixture of sand and salt on roads during and after a snowstorm to improve traction. Using salt brine is more effective than spreading dry salt because moisture is necessary for the freezing-point depression to work and wet salt sticks to the roads better. Otherwise the salt can be wiped away by traffic.

In addition to de-icing, rock salt is occasionally used in agriculture. An example of this would be inducing salt stress to suppress the growth of annual meadow grass in turf production. Other examples involve exposing weeds to salt water to dehydrate and kill them preventing them from affecting other plants. Salt is also used as a household cleaning product. Its coarse nature allows for its use in various cleaning scenarios including grease/oil removal, stain removal, and even dries out and hardens sticky spills for an easier clean

# FLINT

Flint is a hard, sedimentary cryptocrystalline form of the mineral quartz, categorized as the variety of chert that occurs in chalk or marly limestone. Flint was widely used historically to make stone tools and start fires.

It occurs chiefly as nodules and masses in sedimentary rocks, such as chalks and limestones. Inside the nodule, flint is usually dark grey, black, green, white or brown in colour, and often has a glassy or waxy appearance. A thin layer on the outside of the nodules is usually different in colour, typically white and rough in texture. The nodules can often be found along streams and beaches.

Flint breaks and chips into sharp edged pieces, making it useful for knife blades and other cutting tools. The use of flint to make stone tools dates back millions of years, and flint's extreme durability has made it possible to accurately date its use over this time. Flint is one of the primary materials used to define the Stone Age.

During the Stone Age, access to flint was so important for survival that people would travel or trade to obtain flint. Flint Ridge in Ohio was an important source of flint and Native Americans extracted the flint from hundreds of quarries along the ridge. This "Ohio Flint" was traded across the eastern United States and has been found as far west as the Rocky Mountains and south around the Gulf of Mexico.

When struck against steel, flint will produce enough sparks to ignite a fire with the correct tinder, or gunpowder used in weapons. Although it has been superseded in these uses by different processes (the percussion cap), or materials, (ferrocerium), "flint" has lent its name as generic term for a fire starter.

Pebble beach made up of flint nodules eroded out of the nearby chalk cliffs,
Cape Arkona, Rügen, northeast Germany.

The exact mode of formation of flint is not yet clear, but it is thought that it occurs as a result of chemical changes in compressed sedimentary rock formations, during the process of diagenesis. One hypothesis is that a gelatinous material fills cavities in the sediment, such as holes bored by crustaceans or molluscs and that this becomes silicified. This hypothesis certainly explains the complex shapes of flint nodules that are found. The source of dissolved silica in the porous media could be the spicules of silicious sponges. Certain types of flint, such as that from the south coast of England, contain trapped fossilised marine flora. Pieces of coral and vegetation have been found preserved like amber inside the flint. Thin slices of the stone often reveal this effect.

Flint sometimes occurs in large flint fields in Jurassic or Cretaceous beds, for example, in Europe. Puzzling giant flint formations known as paramoudra and flint circles are found around Europe but especially in Norfolk, England on the beaches at Beeston Bump and West Runton.

The "Ohio flint" is the official gemstone of Ohio state. It is formed from limey debris that was deposited at the bottom of inland Paleozoic seas hundreds of millions of years ago that hardened into limestone and later became infused with silica. The flint from Flint Ridge is found in many hues like red, green, pink, blue, white and gray, with the color variations caused by minute impurities of iron compounds.

Flint can be coloured: sandy brown, medium to dark gray, black, reddish brown or an off-white grey.

# Uses

## Tools or Cutting Edges

Neolithic flint axe, about 31 cm long.

Flint was used in the manufacture of tools during the Stone Age as it splits into thin, sharp splinters called flakes or blades (depending on the shape) when struck by another hard object (such as a hammerstone made of another material). This process is referred to as knapping. The process of making tools this way is called "flintknapping".

Flint mining is attested since the Palaeolithic, 3,300,000 years ago, but became more common since the Neolithic (Michelsberg culture, Funnelbeaker culture). In Europe, some of the best tool-making flint has come from Belgium (Obourg, flint mines of Spiennes), the coastal chalks of the English Channel, the Paris Basin, Thy in Jutland (flint mine at Hov), the Sennonian deposits of Rügen, Grimes Graves in England, the Upper Cretaceous chalk formation of Dobruja and the lower Danube (Balkan flint), the Cenomanian chalky marl formation of the Moldavian Plateau (Miorcani flint) and the Jurassic deposits of the Kraków area and Krzemionki in Poland, as well as of the Lägern (silex) in the Jura Mountains of Switzerland.

In 1938, a project of the Ohio Historical Society, under the leadership of H. Holmes Ellis began to study the knapping methods and techniques of Native Americans. Like past studies, this work involved experimenting with actual knapping techniques by creation of stone tools through the use of techniques like direct freehand percussion, freehand pressure and pressure using a rest. Other scholars who have conducted similar experiments and studies include William Henry Holmes, Alonzo W. Pond, Francis H. S. Knowles and Don Crabtree.

To combat fragmentation, flint/chert may be heat-treated, being slowly brought up to a temperature of 150 to 260 °C (300 to 500 °F) for 24 hours, then slowly cooled to room temperature. This makes the material more homogeneous and thus more knappable and produces tools with a cleaner, sharper cutting edge. Heat treating was known to stone age artisans.

## To Ignite Fire or Gunpowder

When struck against steel, a flint edge produces sparks. The hard flint edge shaves off a particle of the steel that exposes iron, which reacts with oxygen from the atmosphere and can ignite the proper tinder.

Prior to the wide availability of steel, rocks of pyrite ($FeS_2$) would be used along with the flint, in a similar (but more time-consuming) way. These methods remain popular in woodcraft, bushcraft, and amongst people practising traditional fire-starting skills.

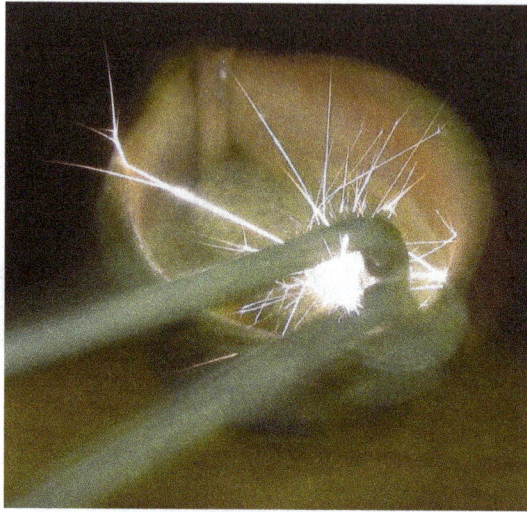

A flint spark lighter in action.

## Flintlocks

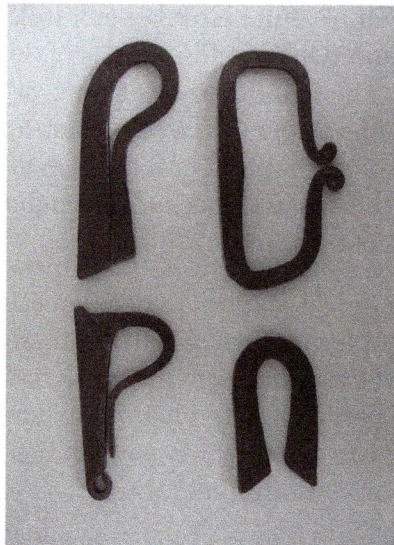

Assorted reproduction firesteels typical of Roman to Medieval period.

A later, major use of flint and steel was in the flintlock mechanism, used primarily in flintlock fire-arms, but also used on dedicated fire-starting tools. A piece of flint held in the jaws of a spring-load-ed hammer, when released by a trigger, strikes a hinged piece of steel ("frizzen") at an angle, cre-ating a shower of sparks and exposing a charge of priming powder. The sparks ignite the priming powder and that flame, in turn, ignites the main charge, propelling the ball, bullet, or shot through the barrel. While the military use of the flintlock declined after the adoption of the percussion cap from the 1840s onward, flintlock rifles and shotguns remain in use amongst recreational shooters.

## Comparison with Ferrocerium

Flint and steel used to strike sparks were superseded by ferrocerium (sometimes referred to as "flint", although not true flint, "mischmetal", "hot spark", "metal match", or "fire steel"). This man-made material, when scraped with any hard, sharp edge, produces sparks that are much hotter than obtained with natural flint and steel, allowing use of a wider range of tinders. Because it can produce sparks when wet and can start fires when used correctly, ferrocerium is commonly included in survival kits. Ferrocerium is used in many cigarette lighters, where it is referred to as "a flint".

## Fragmentation

Flint's utility as a fire starter is hampered by its property of uneven expansion under heating, causing it to fracture, sometimes violently, during heating. This tendency is enhanced by the impurities found in most samples of flint that may expand to a greater or lesser degree than the surrounding stone, and is similar to the tendency of glass to shatter when exposed to heat, and can become a drawback when flint is used as a building material.

## Building Material

Flint, knapped or unknapped, has been used from antiquity (for example at the Late Roman fort of Burgh Castle in Norfolk) up to the present day as a material for building stone walls, using lime mortar, and often combined with other available stone or brick rubble. It was most common in parts of southern England, where no good building stone was available locally, and brick-making not widespread until the later Middle Ages. It is especially associated with East Anglia, but also used in chalky areas stretching through Hampshire, Sussex, Surrey and Kent to Somerset.

Flint was used in the construction of many churches, houses, and other buildings, for example the large stronghold of Framlingham Castle. Many different decorative effects have been achieved by using different types of knapping or arrangement and combinations with stone (flushwork), especially in the 15th and early 16th centuries.

A flint church – the Parish Church of Saint Thomas, in Cricket Saint Thomas, Somerset, England. The height of the very neatly knapped flints varies between 3 and 5 inches (7.6 and 12.7 cm).

Close-up of the wall of the Roman shore fort at Burgh Castle, Norfolk, showing alternating courses of flint and brick.

A typical medieval wall (with modern memorial) at Canterbury Cathedral – knapped and unknapped ("cobble") flints are mixed with pieces of brick and other stones.

Ruins of Thetford Priory show flints and mortar through the whole depth of the wall.

## Ceramics

Flint pebbles are used as the media in ball mills to grind glazes and other raw materials for the ceramics industry. The pebbles are hand-selected based on colour; those having a tint of red, indicating high iron content, are discarded. The remaining blue-grey stones have a low content of chromophoric oxides and so are less deleterious to the colour of the ceramic composition after firing.

Until recently flint was also an important raw material in clay-based ceramic bodies produced in the UK. In preparation for use flint pebbles, frequently sourced from the coasts of South-East England or Western France, were calcined to around 1,000 °C. This heat process both removed organic impurities and induced certain physical reactions, including converting some of the silica to cristobalite. After calcination the flint pebbles were milled to a fine particle size. However, the use of flint has now been superseded by quartz. Because of the historical use of flint, the word "flint" is used by some potters, especially in the US, to refer to siliceous materials that are not flint.

# IRON-RICH SEDIMENTARY ROCKS

Iron-rich sedimentary rocks are sedimentary rocks which contain 15% or more iron. However, most sedimentary rocks contain iron in varying degrees. The majority of these rocks were deposited during specific geologic time periods: The Precambrian (3800 to 570 million years ago), the early Paleozoic (570 to 410 million years ago), and the middle to late Mesozoic (205 to 66 million years ago). Overall, they make up a very small portion of the total sedimentary record.

Iron-rich sedimentary rocks have economic uses as iron ores. Iron deposits have been located on all major continents with the exception of Antarctica. They are a major source of iron and are mined for commercial use. The main iron ores are from the oxide group consisting of hematite, goethite, and magnetite. The carbonate siderite is also typically mined. A productive belt of iron formations is known as an *iron range*.

# Classification

The accepted classification scheme for iron-rich sedimentary rocks is to divide them into two sections: ironstones and iron formations.

## Ironstones

Ironstones consist of 15% iron or more in composition. This is necessary for the rock to even be considered an *iron-rich* sedimentary rock. Generally, they are from the Phanerozoic which means that they range in age from the present to 540 million years ago. They can contain iron minerals from the following groups: oxides, carbonates, and silicates. Some examples of minerals in iron-rich rocks containing oxides are limonite, hematite, and magnetite. An example of a mineral in iron-rich rock containing carbonates is siderite and an example of minerals in an iron-rich rock containing silicate is chamosite. They are often interbedded with limestones, shales, and fine-grained sandstones. They are typically nonbanded, however they can be very coarsely banded on occasion. They are hard and noncherty. The components of the rock range in size from sand to mud, but do not contain a lot of silica. They are also more aluminous. They are not laminated and sometimes contain *ooids*. Ooids can be a distinct characteristic though they are not normally a main component of ironstones. Within ironstones, ooids are made up of iron silicates and/or iron oxides and sometimes occur in alternating laminae. They normally contain fossil debris and sometimes the fossils are partly or entirely replaced by iron minerals. A good example of this is pyritization. They are smaller in size and less likely to be deformed or metamorphosed than iron formations. The term *iron ball* is occasionally used to describe an ironstone nodule.

## Iron Formations

Iron formations must be at least 15% iron in composition, just like ironstones and all iron-rich sedimentary rocks. However, iron formations are mainly Precambrian in age which means that they are 4600 to 590 million years old. They are much older than ironstones. They tend to be cherty, though chert can not be used as a way to classify iron formations because it is a common component in many types of rocks. They are well banded and the banding can be anywhere from a few millimeters to tens of meters thick. The layers have very distinct banded successions that are made up of iron rich layers that alternate with layers of chert. Iron formations are often associates with dolomite, quartz-rich sandstone, and black shale. They sometimes grade locally into chert or dolomite. They can have many different textures that resemble limestone. Some of these textures are micritic, pelleted, intraclastic, peloidal, oolitic, pisolitic, and stromatolitic. In low-grade iron formations, there are different dominant minerals dependent on the different types of facies. The dominant minerals in the oxide facies are magnetite and hematite. The dominant minerals in the silicate facies are greenalite, minnesotaite, and glauconite. The dominant mineral in the carbonate facies is siderite. The dominant mineral in the sulfide facies is pyrite. Most iron formations are deformed or metamorphosed simply due to their incredibly old age, but they still retain their unique distinctive chemical composition; even at high metamorphic grades. The higher the grade, the more metamorphosed it is. Low grade rocks may only be compacted while high grade rocks often can not be identified. They often contain a mixture of banded iron formations and granular iron formations. Iron formations can be divided into subdivisions known as: banded iron formations (BIFs) and granular iron formations (GIFs).

Bog ore.

The above classification scheme is the most commonly used and accepted, though sometimes an older system is used which divides iron-rich sedimentary rocks into three categories: *bog iron deposits*, *ironstones*, and *iron formations*. A bog-iron deposit is iron that formed in a bog or swamp through the process of oxidation.

## Banded Iron Formations vs. Granular Iron Formations

Banded iron formation, North America.

Banded iron formation close up, Upper Michigan.

## Banded Iron Formations

Banded iron formations (BIFs) were originally chemical muds and contain well developed thin lamination. They are able to have this lamination due to the lack of burrowers in the Precambrian. BIFs show regular alternating layers that are rich in iron and chert that range in thickness from a few millimeters to a few centimeters. The formation can continue uninterrupted for tens to hundreds of meters stratigraphically. These formations can contain sedimentary structures like cross-bedding, graded bedding, load casts, ripple marks, mud cracks, and erosion channels. In comparison to GIFs, BIFs contain a much larger spectrum of iron minerals, have more reduced facies, and are more abundant.

## Granular Iron Formations

Granular iron formations (GIFs) were originally well-sorted chemical sands. They lack even, continuous bedding that takes the form of discontinuous layers. Discontinuous layers likely represent bedforms that were generated by storm waves and currents. Any layers that are thicker than a few meters and are uninterrupted, are rare for GIFs. They contain sand-sized clasts and a finer grained matrix, and generally belong to the oxide or silicate mineral facies. Iron formations are sometimes divided into Raptian-type, Algoma-type and Superior-type.

## Algoma Type

Algoma types are small lenticular iron deposits that are associated with volcanic rocks and turbidites. Iron content in this class type rarely exceeds $10^{10}$ tons. They range in thickness from 10-100 meters. Deposition occurs in island arc/back arc basins and intracratonic rift zones.

## Superior Type

Superior types are large, thick, extensive iron deposits across stable shelves and in broad basins. Total iron content in this class type exceeds $10^{13}$ tons. They can extend to over $10^5$ kilometers$^2$. Deposition occurs in relatively shallow marine conditions under transgressing seas.

## Depositional Environment

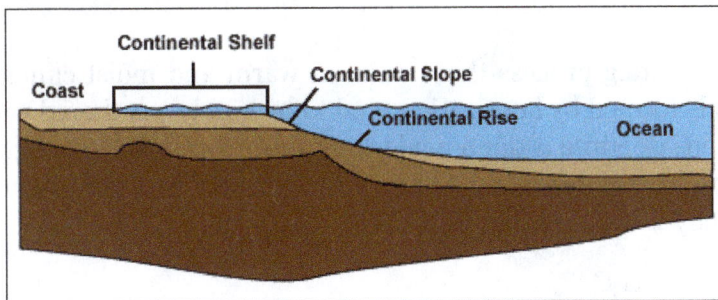

Profile illustrating the shelf, slope and rise.

There are four facies types associated with iron-rich sedimentary rocks: oxide-, silicate-, carbonate-, and sulfide-facies. These facies correspond to water depth in a marine environment. Oxide-facies are precipitated under the most oxidizing conditions. Silicate- and carbonate-facies are precipitated under intermediate redox conditions. Sulfide-facies are precipitated under the most reducing conditions. There is a lack of iron-rich sedimentary rocks in shallow waters which leads to the conclusion that the depositional environment ranges from the continental shelf and upper continental slope to the abyssal plain. (The diagram does not have the abyssal plain labeled, but this would be located to the far right of the diagram at the bottom of the ocean).

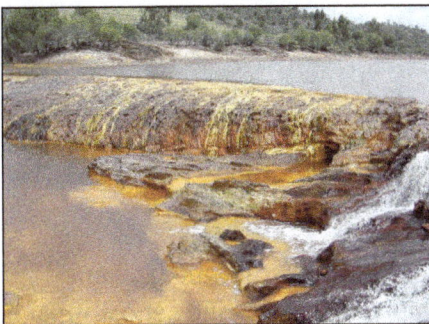

Water colored by oxidized iron, Rio Tinto, Spain.

Iron bacteria growing on iron-rich water seeping.

## Chemical Reactions

Ferrous and ferric iron are components in many minerals, especially within sandstones. $Fe^{2+}$ is in clay, carbonates, sulfides, and is even within feldspars in small amounts. $Fe^{3+}$ is in oxides, hydrous, anhydrous, and in glauconites. Commonly, the presence of iron is determined to be within a rock due to certain colorations from oxidation. Oxidation is the loss of electrons from an element. Oxidation can occur from bacteria or by chemical oxidation. This often happens when ferrous ions come into contact with water (due to dissolved oxygen within surface waters) and a water-mineral reaction occurs. The formula for the oxidation/reduction of iron is:

$$Fe^{2+} \leftrightarrow Fe^{3+} + e^-$$

The formula works for oxidation to the right or reduction to the left.

$Fe^{2+}$ is the ferrous form of iron. This form of iron gives up electrons easily and is a mild reducing agent. These compounds are more soluble because they are more mobile. $Fe^{3+}$ is the ferric form of iron. This form of iron is very stable structurally because it's valence electron shell is half filled.

## Laterization

Laterization is a soil forming process that occurs in warm and moist climates under broadleaf evergreen forests. Soils formed by laterization tend to be highly weathered with high iron and aluminium oxide content. Goethite is often made from this process and is a major source of iron in sediments. However, once it is deposited it must be dehydrated in order to come to an equilibrium with hematite. The dehydration reaction is:

$$2HFeO_2 \rightarrow Fe_2O_3 + H_2O$$

Pyritized Lytoceras.

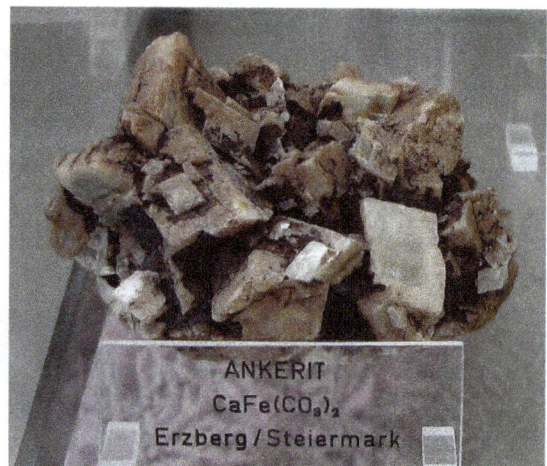

Ankerite, Mineralogical Museum, Bonn, Germany.

## Pyritization

Pyritization is discriminatory. It rarely happens to soft tissue organisms and aragonitic fossils are more susceptible to it than calcite fossils. It commonly takes place in marine depositional environments where there is organic material. The process is caused by sulfate reduction which replaces carbonate skeletons (or shells) with pyrite ($FeS_2$). It generally does not preserve detail and the pyrite forms within the structure as many microcrystals. In fresh water environments, siderite will replace carbonate shells instead of pyrite due to the low amounts of sulfate. The amount of pyritization that has taken place within a fossil may sometimes be referred to as degree of pyritization (DOP).

## Iron Minerals

- Ankerite ($Ca(Mg,Fe)(CO_3)_2$) and siderite ($FeCO_3$) are carbonates and favor alkaline, reducing conditions. They commonly occur as concretions in mudstones and siltstones.

- Pyrite and marcasite ($FeS_2$) are sulfide minerals and favor reducing conditions. They are the most common in fine-grained, dark colored mudstones.

- Hematite ($Fe_2O_3$) is usually the pigment in red beds and requires oxidizing conditions.

- Limonite ($2Fe_2O_3 \cdot 3H_2O$) is used for unidentified massive hydroxides and oxides of iron.

Bournonite with a pyrite crystal matrix, Chichibu Mine, Nakatsugawa, Honshu Island, Japan.

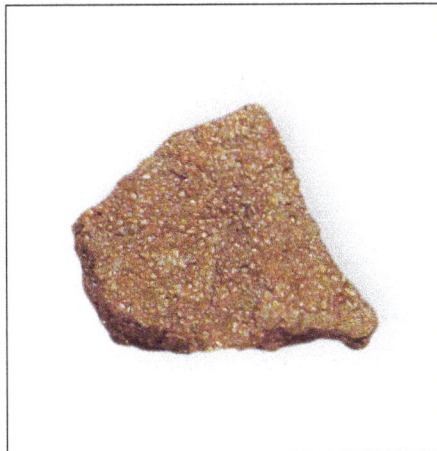

Oolitic Hematite, Clinton, Oneida County, NY.

Limonite, USGS.

## Iron–Rich Rocks in Thin Section

Thin section of rhyolite volcanic rock showing an oxidized iron matrix (orange/brown color).

Magnetite and hematite are opaque under the microscope under transmitted light. Under reflected light, magnetite shows up as metallic and a silver or black color. Hematite will be a more reddish-yellow color. Pyrite is seen as opaque, a yellow-gold color, and metallic. Chamosite is an olive-green color in thin section that readily oxidizes to limonite. When it is partially or fully oxidized to limonite, the green color becomes a yellow-ish brown. Limonite is opaque under the microscope as well. Chamosite is an iron silicate and it has a birefringence of almost zero. Siderite is an iron carbonate and it has a very high birefringence. The thin sections often reveal marine fauna within oolitic ironstones. In older samples, the ooids may be squished and have hooked tails on either end due to compaction.

# CHERT

Chert is a hard, fine-grained sedimentary rock composed of crystals of quartz (silica) that are very small (microcrystalline or cryptocrystalline). Quartz (silica) is the mineral form of silicon dioxide ($SiO_2$). Chert is often of biological origin (organic) but may also occur inorganically as a chemical precipitate or a diagenetic replacement (e.g., petrified wood). Geologists use chert as a generic name for any type of microcrystalline or cryptocrystalline quartz.

Chert is usually of biological origin, being the petrified remains of siliceous ooze, the biogenic sediment that covers large areas of the deep ocean floor, and which contains the silicon skeletal remains of diatoms, silicoflagellates, and radiolarians. Depending on its origin, it can contain either microfossils, small macrofossils, or both. It varies greatly in color (from white to black), but most often manifests as gray, brown, grayish brown and light green to rusty red (occasionally dark green too); its color is an expression of trace elements present in the rock, and both red and green are most often related to traces of iron (in its oxidized and reduced forms respectively).

## Occurrence

Flint with white weathered crust.

Chert occurs in carbonate rocks as oval to irregular nodules in greensand, limestone, chalk, and dolomite formations as a replacement mineral, where it is formed as a result of some type of diagenesis. Where it occurs in chalk or marl, it is usually called flint. It also occurs in thin beds, when it is a primary deposit (such as with many jaspers and radiolarites). Thick beds of chert occur in deep marine deposits. These thickly bedded cherts include the novaculite of the Ouachita Mountains of Arkansas, Oklahoma, and similar occurrences in Texas and South Carolina in the United States. The banded iron formations of Precambrian age are composed of alternating layers of chert and iron oxides.

Chert also occurs in diatomaceous deposits and is known as diatomaceous chert. Diatomaceous chert consists of beds and lenses of diatomite which were converted during diagenesis into dense, hard chert. Beds of marine diatomaceous chert comprising strata several hundred meters thick have been reported from sedimentary sequences such as the Miocene Monterey Formation of California and occur in rocks as old as the Cretaceous.

## Fossils

Chert layer (prominent band near top of outcrop) in the Eocene Ping Tau Formation, Hong Kong.

The cryptocrystalline nature of chert, combined with its above average ability to resist weathering, recrystallization and metamorphism has made it an ideal rock for preservation of early life forms.

For example:

- The 3.2 Ga chert of the figure Tree Formation in the Barbeton Mountains between Swaziland and South Africa preserved non-colonial unicellular bacteria-like fossils.

- The Gunflint Chert of western Ontario (1.9 to 2.3 Ga) preserves not only bacteria and cyanobacteria but also organisms believed to be ammonia-consuming and some that resemble green algae and fungus-like organisms.

- The Apex Chert (3.4 Ga) of the Pilbara craton, Australia preserved eleven taxa of prokaryotes.

- The Bitter Springs Formation of the Amadeus Basin, Central Australia, preserves 850 Ma cyanobacteria and algae.

- The Rhynie chert (410 Ma) of Scotland has remains of a Devonian land flora and fauna with preservation so perfect that it allows cellular studies of the fossils.

## Tools

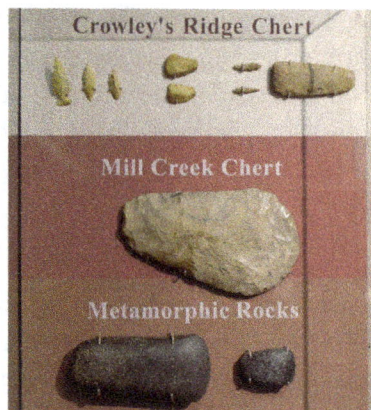

Mill Creek chert from the Parkin Site in Arkansas.

In prehistoric times, chert was often used as a raw material for the construction of stone tools. Like obsidian, as well as some rhyolites, felsites, quartzites, and other tool stones used in lithic reduction, chert fractures in a Hertzian cone when struck with sufficient force. This results in conchoidal fractures, a characteristic of all minerals with no cleavage planes. In this kind of fracture, a cone of force propagates through the material from the point of impact, eventually removing a full or partial cone; this result is familiar to anyone who has seen what happens to a plate-glass window when struck by a small object, such as an air gun projectile. The partial Hertzian cones produced during lithic reduction are called flakes, and exhibit features characteristic of this sort of breakage, including striking platforms, bulbs of force, and occasionally eraillures, which are small secondary flakes detached from the flake's bulb of force.

When a chert stone is struck against an iron-bearing surface, sparks result. This makes chert an excellent tool for starting fires, and both flint and common chert were used in various types of fire-starting tools, such as tinderboxes, throughout history. A primary historic use of common chert and flint was for flintlock firearms, in which the chert striking a metal plate produces a spark that ignites a small reservoir containing black powder, discharging the firearm.

## Construction

Cherts are subject to problems when used as concrete aggregates. Deeply weathered chert develops surface pop-outs when used in concrete that undergoes freezing and thawing because of the high porosity of weathered chert. The other concern is that certain cherts undergo an alkali-silica reaction with high-alkali cements. This reaction leads to cracking and expansion of concrete and ultimately to failure of the material.

In some areas, chert is ubiquitous as stream gravel and fieldstone and is currently used as construction material and road surfacing. Part of chert's popularity in road surfacing or driveway construction is that rain tends to firm and compact chert while other fill often gets muddy when wet.

## Monuments

Chert has been used in late nineteenth-century and early twentieth-century headstones or grave markers in Tennessee and other regions.

## Varieties

There are numerous varieties of chert, classified based on their visible, microscopic and physical characteristics. Some of the more common varieties are:

- Flint is a compact microcrystalline quartz. It was originally the name for chert found in chalk or marly limestone formations formed by a replacement of calcium carbonate with silica. Commonly found as nodules, this variety was often used in past times to make bladed tools. Today, some geologists refer to any dark gray to black chert as flint.

- "Common chert" is a variety of chert which forms in limestone formations by replacement of calcium carbonate with silica. This is the most abundantly found variety of chert. It is generally considered to be less attractive for producing gem stones and bladed tools than flint.

- Jasper is a variety of chert formed as primary deposits, found in or in connection with magmatic formations which owes its red color to iron(III) inclusions. Jasper frequently also occurs in black, yellow or even green (depending on the type of iron it contains). Jasper is usually opaque to near opaque.

- Radiolarite is a variety of chert formed as primary deposits and containing radiolarian microfossils.

- Chalcedony is a microfibrous quartz.

- Agate is distinctly banded chalcedony with successive layers differing in color or value.

- Onyx is a banded agate with layers in parallel lines, often black and white.

- Opal is a hydrated silicon dioxide. It is often of a Neogenic origin. In fact it is not a mineral (it is a mineraloid) and it is generally not considered a variety of chert, although some varieties of opal (opal-C and opal-CT) are microcrystalline and contain much less water (sometime none). Often people without petrological training confuse opal with chert due to similar visible and physical characteristics.

- Magadi-type chert is a variety that forms from a sodium silicate precursor in highly alkaline lakes such as Lake Magadi in Kenya.

- Porcelanite is a term used for fine-grained siliceous rocks with a texture and a fracture resembling those of unglazed porcelain.

- Tripolitic chert (or tripoli) is a light-colored porous friable siliceous (largely chalcedonic) sedimentary rock, which results from the weathering (decalcification) of chert or siliceous limestone.

- Siliceous sinter is porous, low-density, light-colored siliceous rock deposited by waters of hot springs and geysers.

- Mozarkite a varicolored, easily polished Ordovician chert that takes a high polish. It is the state rock of Missouri.

Other lesser used terms for chert (most of them archaic) include firestone, silex, silica stone, chat, and flintstone.

# DOLOMITE

Dolomite (also known as dolostone, dolomite rock or dolomitic rock) is a sedimentary carbonate rock that contains a high percentage of the mineral dolomite, $CaMg(CO_3)_2$. In old USGS publications, it was referred to as magnesian limestone, a term now reserved for magnesium-deficient dolomites or magnesium-rich limestones. Dolomite has a stoichiometric ratio of nearly equal amounts of magnesium and calcium. Most dolomites formed as a magnesium replacement of limestone or lime mud before lithification. Dolomite is resistant to erosion and can either contain bedded layers or be unbedded. It is less soluble than limestone in weakly acidic groundwater, but it can still develop solution features (karst) over time. Dolomite can act as an oil and natural gas reservoir.

The term dolostone was introduced in 1948 to avoid confusion with the mineral dolomite. The usage of the term dolostone is controversial because the name dolomite was first applied to the rock during the late 18th century and thus has technical precedence. The use of the term dolostone is not recommended by the Glossary of Geology published by the American Geological Institute. It is, however, used in some geological publications.

The geological process of conversion of calcite to dolomite is known as dolomitization and any intermediate product is known as "dolomitic limestone."

The "dolomite problem" refers to the vast worldwide depositions of dolomite in the past geologic record eluding a unified explanation for their formation. The first geologist to distinguish dolomite rock from limestone was Belsazar Hacquet in 1778.

## Caves in Dolomite

As with limestone caves, natural caves and solution tubes can form in dolomite rock as a result of dissolution by weak carbonic acid. Calcium carbonate speleothems (secondary deposits) in the forms of stalactites, stalagmites, flowstone etc., can also form in caves within dolomite rock. "Dolomite is a common rock type, but a relatively uncommon mineral in speleothems". Both the 'Union Internationale de Spéléologie' (UIS) and the American 'National Speleological Society' (NSS), extensively use in their publications, the terms "dolomite" or "dolomite rock" when referring to the natural bedrock containing a high percentage of $CaMg(CO_3)_2$ in which natural caves or solution tubes have formed.

## Dolomite Speleothems

Both calcium and magnesium go into solution when dolomite rock is dissolved. The speleothem precipitation sequence is: calcite, Mg-calcite, aragonite, huntite and hydromagnesite. Hence, the most common speleothem (secondary deposit) in caves within dolomite rock karst, is calcium carbonate in the most stable polymorph form of calcite. Speleothem types known to have a dolomite constituent include: coatings, crusts, moonmilk, flowstone, coralloids, powder, spar and rafts. Although there are reports of dolomite speleothems known to exist in a number of cave around the world, they are usually in relatively small quantities and form in very fine-grained deposits.

# CARBONATE ROCK

Carbonate rocks are made of particles (composed >50% carbonate minerals) embedded in a cement. Most carbonate rocks result from the accumulation of bioclasts created by calcareous organisms. Therefore carbonate rocks originate in area favoring biological activity i.e. in shallow and warm seas in areas with little to no siliciclastic input. In present day Earth these areas are limited to ±40 latitude in region away or protected from erosion-prone elevated continental areas.

In the Devonian high-sea level, favoring the development of shallow epicontinental sea, warmer conditions, and relatively low detrital sedimentation rates, resulted in larger thick carbonate buildups. The Murrumbidgee serie is an example of such carbonate buildup.

Because they contain numerous bioclasts (i.e. fossils) carbonate rocks are cherish by stratigraphers and paleontologist alike as they enable them to reconstruct stratigraphic sequences and make large-scale correlation between geological formations hundred to thousand of kilometer apart. In addition carbonate rocks' texture and the nature of their bioclasts offer a detailed insight into their depositional environment.

## Depositional Environments

Carbonate depositional environment past or present fall into three general types:

- Ramp continental margins are continental platform gently sloping toward the ocean (<1). They are limited by the emerged continent and toward the oceanic basin by the gentle continental break. It is an environment of high energy as the amplitude of the waves increases as the depth of the sea decreases.

- Rimmed margins are continental platform limited toward the ocean by a steep and abrupt continental break where a nearly continuous carbonate rim or barrier develops. These rims or barriers are wave resistant structure made of reef coral (alive or dead) and oolitic sand shoals. Landward the rim/barrier is a low energy environment of "lagoonal" characteristics grading landward into a tidal flat. In present Earth they can be recognized by colorful tourists snorkeling happily together. Oceanward, the rim/barrier is a high-energy environment. In this noisy, wavy and scummy environment it takes commitment, strength and a bit of an attitude to any living organism to hold on onto the place. Tourists are rarely seen in those dangerous waters.

- Isolated platforms, also known as "Bahama type", are shallow platform 10's to 100's km wide offshore of shallow continental shelves, surrounded by 100's to 1000's m deep water. This environment is characterized by the absence of siliciclastic input.

## Dunham Classification

The Dunham classification (1962) is based on concept of grain/mud support therefore on the proportion mud-particle and the depositional textures. The concept of "support" assume continuity of either the mud matrix or that of the grains. If the carbonate is mud-supported the grains float into a continuum of mud matrix. In grain-supported carbonate rocks the grains from an interconnected skeleton in which the mud fills the gap. The percentage mud/cement at which there is a switch between mud-supported and grain-supported depends on the fabric (preferential agencement) of the particles.

| Deposition Texture Recognizable | | | | | Unrecognizable |
|---|---|---|---|---|---|
| Original Components Not Bound Together During Deposition | | | | Original Bound | |
| Contain Mud, Clay, And Fine Silt-Size Carbonate | | | No Mud | | |
| Mud Supported | | Grain-Supprtd | | | |
| Grains < 10% | Grains > 10% | | | | |
| | | | | | |
| Mudstones | Wakestone | Packstone | Grainstone | Boundstone | Crystalline |

# References

- 6-4-sedimentary-structures-and-fossils, chapter, geology: opentextbc.ca, Retrieved 29 July, 2019

- Monroe, James S., and Reed Wicander. The Changing Earth: Exploring Geology and Evolution, 2nd ed. Belmont: West Publishing Company, 1997. ISBN 0-314-09577-2 pp. 114

- Bedding, geology-and-oceanography, geology-and-oceanography, earth-and-environment: encyclopedia.com, Retrieved 30 August, 2019

- Buatois, Luis A.; Encinas, Alphonso (April 2011). "Ichnology, Sequence Stratigraphy and Depositional Evolution of an Upper Cretaceous Rocky Shoreline in Central Chile: Bioerosion Structures in a Transgressed Metamorphic Basement". Cretaceous Research. 32 (2): 203–212. Doi:10.1016/j.cretres.2010.12.003

- Formation-types-and-examples-of-sedimentary-rocks, geology: eartheclipse.com, Retrieved 1 January, 2019

- "Openlearn Live: 19th February 2016: A Week In South Carolina:". Openlearn. The Open University. Retrieved 20 February 2016

- 6-1-clastic-sedimentary-rocks, chapter, geology: opentextbc.ca, Retrieved 2 February, 2019

- Monroe, James S., and Reed Wicander. The Changing Earth: Exploring Geology and Evolution, 2nd ed. Belmont: West Publishing Company, 1997. ISBN 0-314-09577-2 pp. 114-15, 352

- Shale-rock-4165848: thoughtco.com, Retrieved 3 March, 2019

- Mcbride, E. F. (2003), Pseudofaults resulting from compartmentalized Liesegang bands: update. Sedimentology, 50: 725–730. Doi:10.1046/j.1365-3091.2003.00572

- Carbonate, Yass04, fieldtrips, prey, users: geosci.usyd.edu.au, Retrieved 4 April, 2019

- Prothero, Donald R.; Schwab, Fred (22 August 2003). Sedimentary Geology. Macmillan. Pp. 265–269. ISBN 978-0-7167-3905-0. Retrieved 15 December 2012

# 5

# Sedimentary Basin

Sedimentary basin refers to a depression in the Earth's crust that is formed by plate tectonic activity in which sediments accumulate. They can be classified on the basis of the tectonic setting and geometry palaeography. The chapter closely examines the classifications of sedimentary basins to provide an extensive understanding of the subject.

Sedimentary basins are areas in which sediment accumulated at a sig-nificantly greater rate than sediments of the same age in neighbouring areas, so accumulating a greater thickness. The sediments accumulate by virtue of subsidence.

The sedimentary veneer on the Earth's surface varies greatly in thickness. If we stand in central Siberia or south-central Canada, we will find ourselves on igneous and metamorphic basement rocks that are over a billion years old sedimentary rocks are nowhere in sight. Yet if we stand along the southern coast of Texas, we would have to drill through over 15 km of sedimentary beds before reaching igneous and metamorphic basement. Thick accumulations of sediment form only in special regions where the surface of the Earth's lithosphere sinks, providing space in which sediment collects. Geologists use the term subsidence to refer to the process by which the surface of the lithosphere sinks, and the term sedimentary basin for the sediment-filled depression. In what geologic settings do sedimentary basins form? An understanding of plate tectonics theory provides the answers.

## Categories of Basins in the Context of Plate Tectonics Theory

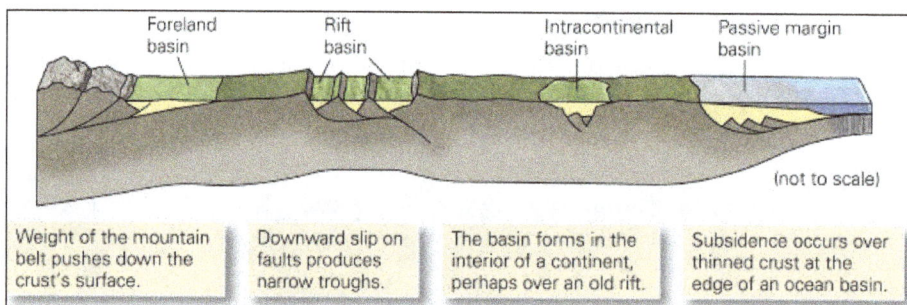

| Foreland basin | Rift basin | Intracontinental basin | Passive margin basin |
|---|---|---|---|
| Weight of the mountain belt pushes down the crust's surface. | Downward slip on faults produces narrow troughs. | The basin forms in the interior of a continent, perhaps over an old rift. | Subsidence occurs over thinned crust at the edge of an ocean basin. |

(not to scale)

The geologic setting of sedimentary basins.

Geologists distinguish among different kinds of sedimentary basins in the context of plate tectonics theory. Let's consider a few examples.

1. Rift basins: These form in continental rifts, regions where the lithosphere is stretching horizontally, and therefore thins vertically. As the rift grows, slip on faults drops blocks of crust down, producing low areas bordered by narrow mountain ridges. These troughs fill with sediment.

2.  Passive-margin basins: These form along the edges of continents that are not plate bound-aries. They are underlain by stretched lithosphere, the remnants of a rift whose evolu-tion successfully led to the formation of a mid-ocean ridge and subsequent growth of a new ocean basin. Passive-margin basins form because subsidence of stretched lithosphere continues long after rifting ceases. They fill with sediment carried to the sea by rivers and with carbonate rocks formed in coastal reefs.

3.  Intracontinental basins: These develop in the interiors of continents, initially because of subsidence over a rift. They may continue to subside in pulses even hundreds of millions of years after they formed, for reasons that are not well understood.

4.  Foreland basins: These form on the continent side of a mountain belt because the forces produced during convergence or collision push large slices of rock up faults and onto the surface of the continent. The weight of these slices pushes down on the surface of the lith-osphere, producing a wedge-shaped depression adjacent to the mountain range that fills with sediment eroded from the range. Fluvial and deltaic strata accumulate in foreland basins.

## Transgression and Regression

Sea-level changes, relative to the land surface, control the succession of sediments that we see in a sedimentary basin. At times during Earth history, sea level has risen by as much as a couple of hundred meters, creating shallow seas that submerge the interiors of continents. At other times, sea level has fallen by a couple of hundred meters, exposing the continental shelves to air. Global sea-level changes may be due to a number of factors, including climate change, which controls the amount of ice stored in polar ice caps and causes changes in the volume of ocean basins. Sea level at a location may also be due to the local uplift or sinking of the land surface.

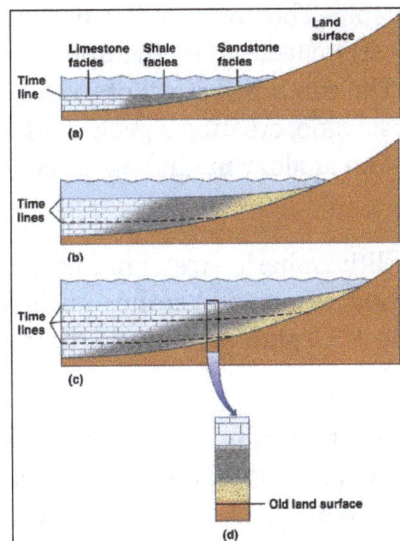

The concept of transgression and regression, during deposition of sedimentary sequence.

When relative sea level rises, the shoreline migrates inland. We call this process transgression. When relative sea level falls, the coast migrates seaward. We call this process regression. The pro-cess of transgression and regression leads to the formation of broad blankets of sediment.

# Diagenesis

Lithification is an aspect of a broader phenomenon called diagenesis. Geologists use the term dia-genesis for all the physical, chemical, and biological processes that transform sediment into sedi-mentary rock and that alter characteristics of sedimentary rock after the rock has formed.

In sedimentary basins, sedimentary rocks may become very deeply buried. As a result, the rocks endure higher pressures and temperatures and come in contact with warm groundwater. Diagen-esis, under such conditions, can cause chemical reactions in the rock that produce new minerals and can also cause cement to dissolve or precipitate.

As temperature and pressure increase still deeper in the subsurface, the changes that take place in rocks become more profound. At sufficiently high temperature and pressure, metamorphism begins, in that a new assemblage of minerals forms, and/or mineral grains become aligned paral-lel to each other. The transition between diagenesis and metamorphism in sedimentary rocks is gradational and occurs between temperatures of 15 °C and 30 °C.

Sedimentary basins form primarily in convergent, divergent and transform settings. Convergent boundaries create foreland basins through tectonic compression of oceanic and continental crust during lithospheric flexure. Tectonic extension at divergent boundaries where continental rifting is occurring can create a nascent ocean basin leading to either an ocean or the failure of the rift zone. In tectonic strike-slip settings, accommodation spaces occur as transpressional, transten-sional or transrotational basins according to the motion of the plates along the fault zone and the local topography pull-apart basins.

## Lithospheric Stretching

If the lithosphere is caused to stretch horizontally, by mechanisms such as *ridge-push* or *trench-pull*, the effect is believed to be twofold. The lower, hotter part of the lithosphere will "flow" slow-ly away from the main area being stretched, whilst the upper, cooler and more brittle crust will tend to fault (crack) and fracture. The combined effect of these two mechanisms is for the Earth's surface in the area of extension to subside, creating a geographical depression which is then often infilled with water and sediments. (An analogy might be a piece of rubber, which thins in the mid-dle when stretched.)

An example of a basin caused by lithospheric stretching is the North Sea – also an important location for significant hydrocarbon reserves. Another such feature is the Basin and Range Province which covers most of the USA state of Nevada, forming a series of horst and graben structures.

Another expression of lithospheric stretching results in the formation of ocean basins with central ridges; The Red Sea is in fact an incipient ocean, in a plate tectonic context. The mouth of the Red Sea is also a tectonic triple junction where the Indian Ocean Ridge, Red Sea Rift and East African Rift meet. This is the only place on the planet where such a triple junction in oceanic crust is exposed subaerially. The reason for this is twofold, due to a high thermal buoyancy of the junction, and a local crumpled zone of seafloor crust acting as a dam against the Red Sea.

## Lithospheric Compression/Shortening and Flexure

If a load is placed on the lithosphere, it will tend to flex in the manner of an elastic plate. The magnitude of the lithospheric flexure is a function of the imposed load and the *flexural rigidity* of the lithosphere, and the wavelength of flexure is a function of flexural rigidity alone. Flexural rigidity is in itself, a function of the lithospheric mineral composition, thermal regime, and effective elastic thickness. The nature of the load is varied. For instance, the Hawaiian Islands chain of volcanic edifices has sufficient massto cause deflection in the lithosphere.

The obduction of one tectonic plate onto another also causes a load and often results in the creation of a foreland basin, such as the Po basin next to the Alps in Italy, the Molasse Basin next to the Alps in Germany, or the Ebro basin next to the Pyrenees in Spain.

## Strike-Slip Deformation

Deformation of the lithosphere in the plane of the earth (i.e. such that faults are vertical) occurs as a result of near horizontal maximum and minimum principal stresses. The resulting zones of subsidence are known as strike-slip or pull-apart basins. Basins formed through strike-slip action occur where a vertical fault plane curves. When the curve in the fault plane moves apart, a region of *transtension* results, creating a basin. Another term for a transtensional basin is a *rhombochasm*. A classic rhombochasm is illustrated by the Dead Sea rift, where northward movement of the Arabian Plate relative to the Anatolian Plate has caused a rhombochasm.

The opposite effect is that of *transpression*, where converging movement of a curved fault plane causes collision of the opposing sides of the fault. An example is the San Bernardino Mountains north of Los Angeles, which result from convergence along a curve in the San Andreas fault system. The Northridge earthquake was caused by vertical movement along local thrust and reverse faults *bunching up* against the bend in the otherwise strike-slip fault environment. In Nigeria, the dominant type of basement rock intersected by wells drilled for hydrocarbons, limestone, or water is granite. The three sedimentary basins in Nigeria are underlain by continental crust except in the Niger delta, where the basement rock is interpreted to be oceanic crust. Most of the wells that penetrated the basement are in the Eastern Dahomey embayment of western Nigeria. A maximum thickness of about 12,000 m of sedimentary rocks is attained in the offshore western Niger delta, but maximum thicknesses of sedimentary rocks are about 2,000 m in the Chad basin and only 500 m in the Sokoto embayment.

# FORMATION OF SEDIMENTARY BASIN

Sedimentary basins range in size from as small as hundreds of meters to large parts of ocean basins. The essential element of the concept is tectonic creation of relief, to provide both a source of sediment and a relatively low place for the deposition of that sediment. Tectonics is the most important control on sedimentation; climate is a rather distant second. The important effects of tectonics on sedimentation, direct or indirect, include the following:

- Nature of sediment

- Rate of sediment supply

- Rate of deposition

- Depositional environment

- Nature of source rocks

- Nature of vertical succession

The only basins that are preserved in their entirety are those that lie entirely in the subsurface! Basins exposed at the surface are undergoing destruction and loss of record by erosion. So there's an ironic trade-off between having more complete preservation in the subsurface but less satisfactory observations.

- Master cross sections: With the present land surface as the most natural datum, construct several detailed physical cross sections through the basin to show its geometry and sediment fill.

- Stratigraphic sections: Construct a graph, with time along the vertical axis, showing the time correlations of all the major rock units along some generalized traverse across the basin. Such a section includes hiatuses, during which there was non deposition or erosion.

- Isopach maps: With some distinctive stratigraphic horizon near the top of the section as datum, draw a contour map showing isopachs (isopachs are loci of equal total sediment thickness) in the basin.

- Lithofacies maps: For one or a series of times, draw a map showing distribution of sediment types being deposited at that time.

- Ratio maps: Compute things like sand/shale ratio, integrated over the entire section or restricted to some time interval, and plot a contour map of the values.

- Paleocurrent maps: For one or a series of times, draw a map showing the direction of paleo-currents in the basin at that time.

- Grain-size maps: For the entire basin fill, averaged vertically, or for some stratigraphic interval or time interval, draw a map that shows the areal distribution of sediment grain size. This is especially useful for conglomeratic basins.

In one sense, the origin of sedimentary basins boils down to the question of how relief on the Earth is created. Basically, there are only a few ways, described in the following sections.

## Local

On a small scale, hundreds to thousands of meters laterally, fault movements can create relief of hundreds to thousands of meters, resulting in small but often deep basins (some of these are called intermontane basins). Along strike-slip faults. can produce small pull-apart basins; more on them later. Relief of this kind is on such a small scale that it tends not to be isostatically compensated. It's like setting a block of granite out on your driveway; the flexural rigidity of your driveway is great enough compared with the imposed load that the granite block is prevented from finding its buoyant equilibrium position.

## Regional

Basin relief can be created mechanically on a regional scale in two very important ways: thermally or flexurally, or by a combination of those two effects). Each of these is discussed briefly below. Keep in mind that basins can also be made just by making mountain ranges, on land or in the ocean, by volcanism.

## Thermal

If the lithosphere is heated from below, it expands slightly and thus becomes less dense. This less dense lithosphere adjusts isostatically to float higher in the asthenosphere, producing what we see at the Earth's surface as crustal uplift. If the lithosphere cools back to its original temperature, there's isostatic subsidence back to the original level.

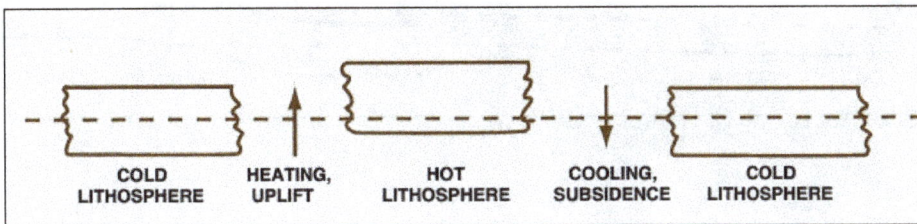

But suppose that some erosion took place while the crust was elevated. The crust is thinned where the erosion took place (and thickened somewhere else, where there was deposition; that might be far away, at the mouth of some long river system), so when the crust cools again it subsides to a position lower than where it started, thus creating a basin available for filling by sediments.

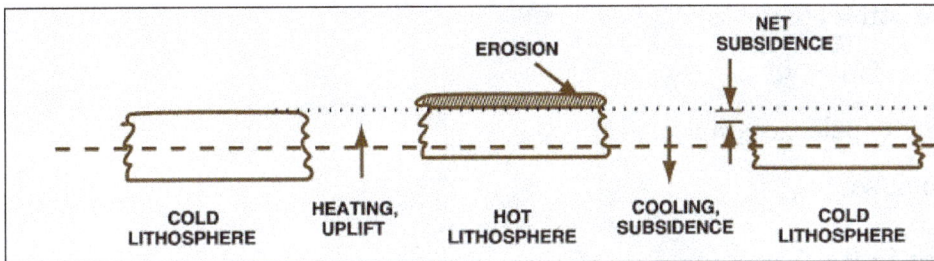

But the magnitude of crustal lowering by this mechanism is less than is often observed in basins thought to be created thermally. It has therefore been proposed, and widely accepted, that in many cases extensional thinning of the lithosphere accompanies the heating. Then, upon re-cooling, the elevation of the top of the lithosphere is less than before the heating and extension. This kind of subsidence has been invoked to explain many sedimentary basins.

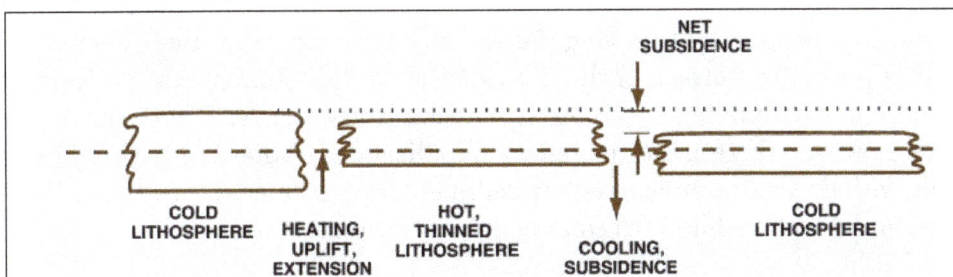

## Flexural

Another important way to make basins is to park a large load on some area of the lithosphere. The new load causes that lithosphere to subside by isostatic adjustment. But because the lithosphere has considerable flexural rigidity, adjacent lithosphere is bowed down also. The region between the high-standing load and the lithosphere in the far field (in the parlance of geophysics, that just means far away!) is thus depressed to form a basin. This model has been very successful in accounting for the features of foreland basins, which are formed ahead of large thrust sheets that move out from orogenic areas onto previously undeformed cratonal lithosphere.

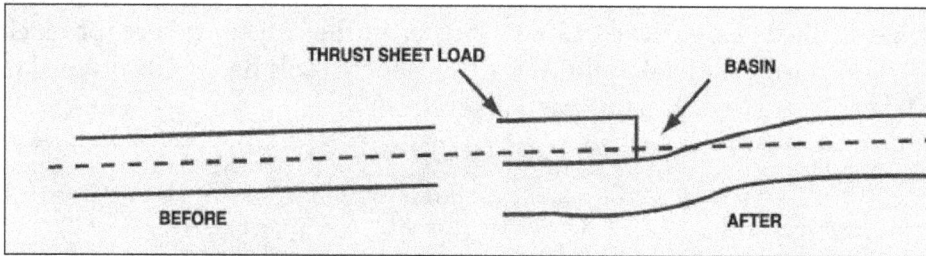

# CLASSIFICATION OF SEDIMENTARY BASINS

Sedimentary basins are classified by tectonic (and, specifically, plate-tectonic) setting. That's fairly easy to do for modern basins, but it's rather difficult to do for ancient basins. List of some of the important criteria that could be used, ranging from more descriptive at the top of the list to more genetic at the bottom of the list:

- Nature of fill
- Geometry paleogeography
- Tectonic setting

## Intracratonic Basins

Location and tectonic setting: in anorogenic areas on cratons.

Tectonic and sedimentary processes: These basins have no apparent connection with plate tectonics. They are thought to reflect very slow thermal subsidence (for times of the order of a hundred million years) after a heating event under the continental lithosphere. But the reasons for depression below the original crustal level are not understood. Erosion during the thermal uplift seems untenable, as does lithospheric stretching. Subsidence is so slow that there seems to have been no depression of the upper surface of the lithosphere, so depositional environments are mostly the same as those in surrounding areas; the succession is just thicker. These successions are also more complete, however, there are fewer and smaller diastems (A diastem is a brief interruption in sedimentation, with little or no erosion before sedimentation resumes), so at times the basin must have remained under water while surrounding areas were emergent.

Size, shape: rounded, equidimensional, hundreds of kilometres across.

Sediment fill: shallow-water cratonal sediments (carbonates, shales, sandstones), thicker and more complete than in adjacent areas of the craton but still relatively thin, hundreds of meters.

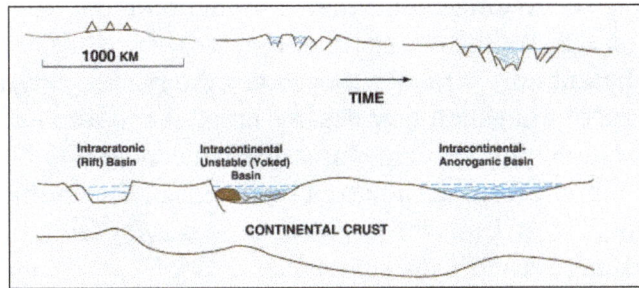

## Aulacogens

Location and tectonic setting: extending from the margins toward the interiors of cratons.

Tectonic and sedimentary processes: Aulacogens are thought to represent the third, failed arm of a three-armed rift, two of whose arms continued to open to form an ocean basin. In modern settings, aulacogens end at the passive continental margin. An example is the Benue Trough, underlying the Congo River Basin in West Africa. The ocean eventually closes to form an orogenic belt, so in ancient settings aulacogens end at an orogenic belt; an example is the basin filled by the Pahrump Group (Proterozoic) in Nevada and California.

Size, shape: long, narrow, linear; tens of kilometres wide, many hundreds of kilometres long.

Sediment fill: very thick (up to several thousand meters); coarse to fine siliciclastics, mostly coarse, minor carbonates; mostly non-marine, some marine; contemporaneous folding and faulting; the succession often passes upward, with or without major unconformity, into thinner and more widespread shallow-marine cratonal siliciclastics and carbonates.

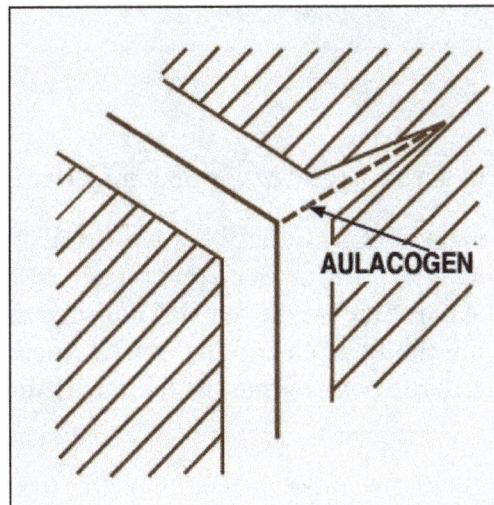

## Rift Basins

Location and tectonic setting: Within continental lithosphere on cratons.

Tectonic and sedimentary processes: Lithospheric extension on a craton, presumably by regional sub-lithospheric heating, causes major rifts. Some such rifts continue to open and eventually become ocean basins floored by oceanic rather than continental crust; the basin description here then applies to this earliest stage of the rifting. In other cases, the rifts fail to open fully into ocean basins (perhaps some adjacent and parallel rift becomes the master rift), so they remain floored by thinned continental crust rather than new oceanic crust. A modern example: the East African rift valleys. An ancient example: the Triassic–Jurassic Connecticut and Newark basins in eastern North America. Sediment supply from the adjacent highlands of the uplifted fault blocks is usually abundant, although in the East African rifts the land slope is away from the rim of the highlands, and surprisingly little sediment reaches the rift basins.

Size, shape: long, narrow, linear; tens of kilometres wide, up to a few thousand kilometres long.

Sediment fill: Coarse to fine siliciclastics, usually non-marine; often lacustrine sediments; inter-bedded basalts.

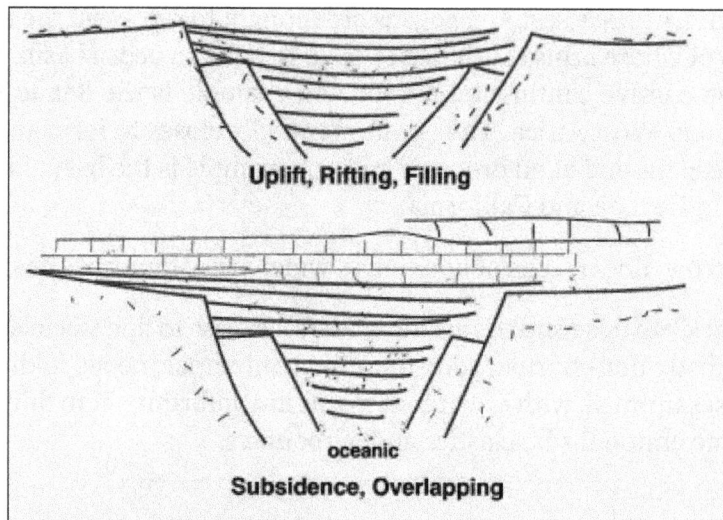

## Oceanic Rift Basins

Location and tectonic setting: In a narrow and newly opening ocean.

Processes: This category of basins is transitional between intracontinental rift basins and passive-margin basins. Basins have opened wide enough to begin to be floored with oceanic crust but are still so narrow that the environment is either still non-marine or, if marine, has restricted circulation. Modern examples are the Red Sea and the Gulf of Aden. In the ancient, the sediment fill of such a basin is likely to underlie passive-margin sediments deposited later in the history of ocean opening.

Size, shape: long, narrow; straight or piecewise straight; tens to a few hundreds of kilometres wide, up to a few thousand kilometres long.

Sediment fill: mafic, volcanics and coarse to fine non-marine siliciclastics, as in intracratonic rift basins described above, passing upward and laterally into evaporites, lacustrine deposits, and fine marine sediments, often metal-rich from hydrothermal activity at the spreading ridge.

Rift Valley Phase
Lavas - Sediments
Potential Erosion
Proto-Oceanic Gulf Phase
MSL
MSL
Full Crust
MOHO
Standard Oceanic Crust
Quasi-oceanic Crust
MOHO
Full Crust
Basal Clastic Phase
MSL
20
15
10
5
0
KILOMETERS
Full Crust
Pre-Oceanic Phases
Transitional crust
Oceanic Crust
Continental Terrace
Carbonate-Shale Shelfal Phase
MSL
Turbidites
Continental Rise
Time
Transitional crust
Full Crust
Oceanic Crust
Continental Embankment
MSL
Transitional crust
Full Crust
Oceanic Crust
0 100 200 300 400 500
KILOMETERS

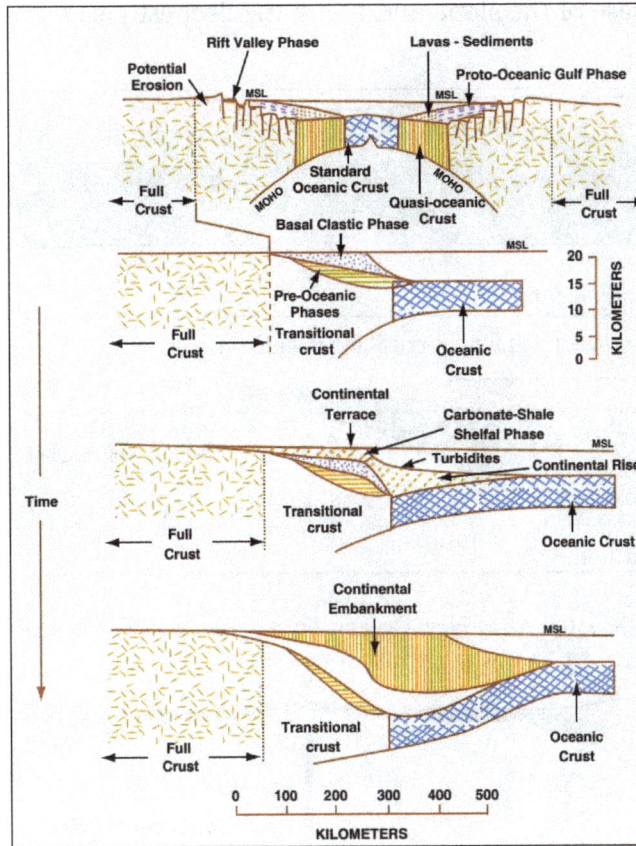

## Passive Margin Basins

Location and tectonic setting: Along passive continental margins, approximately over the transition from continental to oceanic crust formed by rifting and opening of a full-scale ocean basin.

Processes: As an ocean basin opens by spreading, the zone of heating and extensional thinning of continental crust on either side of the ocean basin subsides slowly by cooling. Sediments, either siliciclastics derived from land or carbonates generated in place, cover this subsiding transition from continental crust to oceanic crust with a wedge of sediment to build what we see today as the continental shelf and slope. In the context of the ancient, this represents the miogeocline. The subsidence is accentuated by loading of the deposited sediments, resulting in a prominent down bowing of the continental margin. Deposition itself therefore does not take place in a basinal geometry, but the base of the deposit is distinctively concave upward.

Size, shape: Straight to piecewise straight, often with considerable irregularity in detail; a few hundreds of kilometres wide, thousands of kilometres long.

Sediment fill: Overlying and overlapping the earlier deposits laid down earlier during rifting and initial opening are extensive shallow-marine siliciclastics and carbonates of the continental shelf, thickening seaward. These sediments pass gradually or abruptly into deeper marine fine sediments of the continental slope and rise, often grading or inter-fingering seaward into deep marine coarse and fine siliciclastics or re-sedimented carbonates in the form of turbidites building submarine fans at the base of the slope and filling the deepest parts of the ocean basin to form abyssal plains.

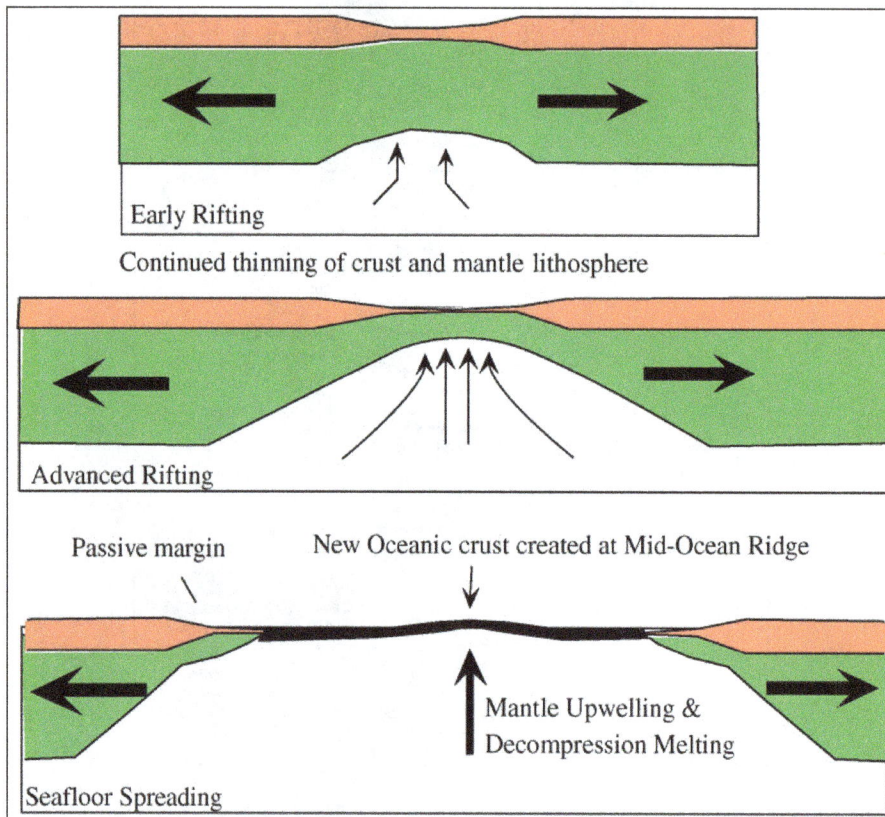

Early Rifting

Continued thinning of crust and mantle lithosphere

Advanced Rifting

Passive margin          New Oceanic crust created at Mid-Ocean Ridge

Mantle Upwelling & Decompression Melting

Seafloor Spreading

## Trenches

Location and tectonic setting: In the abyssal ocean, at the line of initial down bending of the subducted oceanic-crust plate in a subduction zone.

Tectonic and sedimentary processes: (1) Open-ocean pelagic sediments (mainly abyssal brown clay and organic oozes) are conveyor-belted to the trench, and underlie the sediments, thin to thick, deposited in the trench itself. (2) What happens to the sediments delivered to or deposited in the trench? While still within the trench they are little deformed, but they don't stay that way long. They are either scraped off the descending plate to form an accretionary wedge, whose structure ranges from chaotically mixed material in a subduction mélange to a fairly regular imbricated succession of under-thrusted sheets dipping toward the arc (the thrust sheets themselves get younger downward, but within a given plate the sediments get younger upward), or they are dragged down the subduction zone.

Size, shape: long and narrow (tens of kilometres wide, thousands of kilometres long), arcuate, with convex side toward the oncoming subducted plate.

Sediment fill: varies from thin (hundreds of meters) pelagic sediments (fine abyssal muds, volcanic ash) to thick (thousands of meters) arc-derived coarse siliciclastics and volcaniclastics, as local fans built perpendicular to the trench axis or oblong fan like bodies built parallel to trench axis.

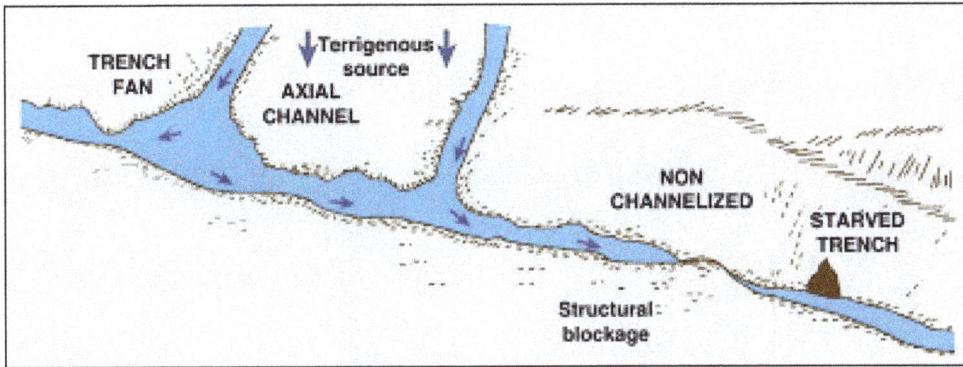

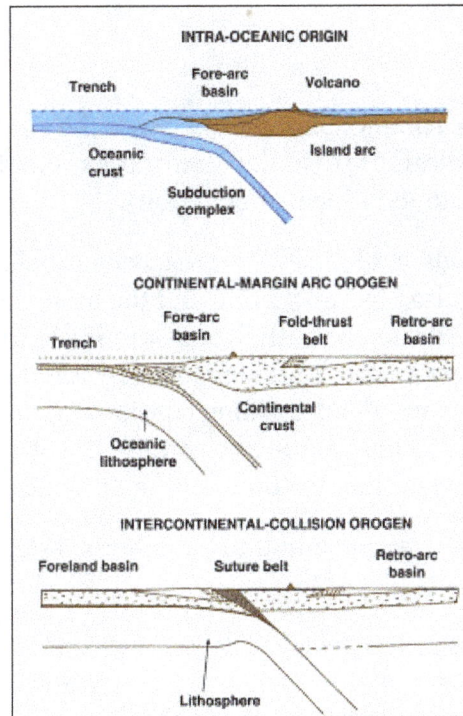

## Trench-Slope Basins

Location and tectonic setting: On the inner (arcward) wall of subduction-zone trenches.

Tectonic and sedimentary processes: The basins are formed as low areas, with closed contours, between adjacent thrust sheets in the growing accretionary wedge. Near-surface folding of sediment in the accretionary wedge may also be a factor in the development of the basins. These basins intercept some of the sediment carried as turbidity currents from upraised older parts of the accretionary complex, or from the more distant arc itself.

Size, shape: Small (no larger than kilometres wide, tens of kilometres long, often smaller); linear, and elongated parallel to the trench.

Sediment fill: deep-marine silts and muds sedimented directly into the basins or slumped into the basins from higher on the slope; also coarser siliciclastics supplied from farther upslope by turbidity currents.

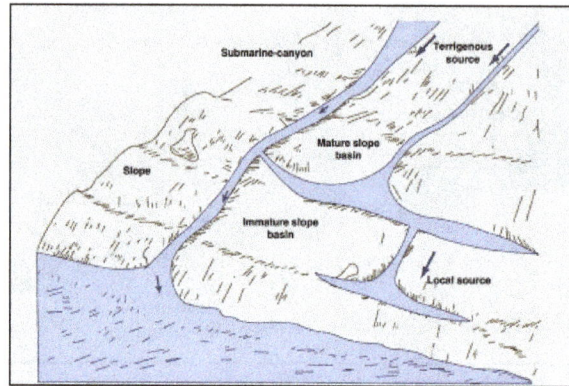

## Fore-Arc Basins

Location and tectonic setting: In subduction zones; between the upraised subduction complex just inboard of the trench and the volcanic arc (in the case of ocean-ocean subduction) or the overriding continent (in the case of ocean–continent subduction).

Processes: As subduction proceeds, a relatively low area, usually below sea level, is formed between the relatively high outer arc upraised by subduction and the inner volcanic arc built by subduction magmatism. Ancient examples of such fore-arc basins are likely to be tectonically isolated from the originally adjacent areas. After an arc-continent collision, another variant of fore-arc basin can be formed between the outer arc and the overriding continent. These basins are likely to be filled mainly from the high land of the tectonically active continent. Later continent-continent collision would make direct reconstruction of their tectonic setting difficult.

Size, shape: tens of kilometres to over one hundred kilometres wide, up to thousands of kilometres long; commonly arcuate.

Sediment fill: non-marine, siliciclastic, fluvial to deltaic deposits at the arcward margin pass seaward into deep marine siliciclastics, mainly sediment-gravity flow deposits, all inter-bedded with arc derived volcanics flows and pyroclastics. Section thickness can be many thousands of meters.

## Foreland Basins

Location, tectonic setting, processes: There are two kinds of foreland basins: retro-arc foreland basins, which are formed on stable continental crust by loading by thrust sheets moving toward the continental interior as a result of compression and crustal shortening in an ocean-continent subduction zone, and peripheral foreland basins, formed after continent-continent collisions by loading of the continental crust of the subducted plate by development of thrust sheets in the continental crust of the subducted plate directed back away from the subduction zone. Both kinds tend to be asymmetrical, with their deepest parts nearest the emplaced thrust sheets. They tend to migrate away from the arc or suture zone with time. They are filled by sediments derived from the mountainous terrain associated with the compression and thrusting.

Size, shape: tens to a few hundreds of kilometres wide, hundreds to thousands of kilometres long; often with varying development along their length; commonly arcuate or piecewise arcuate, reflecting the geometry of subduction.

Sediment fill: Coarse, fluvial, siliciclastics, mainly as alluvial fans, thinning and fining away from the arc or suture, often passing into shallow-marine sandstone-shale successions if sea level is high enough to flood the basin. Thickness are up to many thousands of meters. The classic molasse facies, thick non-marine conglomerates, is deposited in foreland basins.

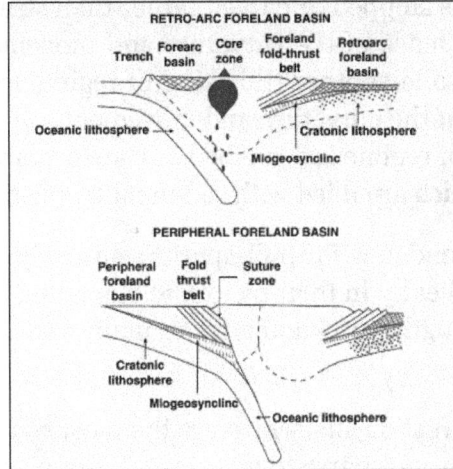

## Remnant Basins

Location and tectonic setting: within suture zones formed by continent-continent collision.

Processes: Continental margins and subduction zones are (for different reasons, connected with geometry of rifting and geometry of subduction) commonly irregular in plan rather than straight, so when continent-continent collision eventually comes to pass, certain salient of continental crust encounter the subduction zone before re-entrants. With further subduction and suturing, this creates isolated basins still floored by residual oceanic crust, which receive abundant sediment from adjacent strongly uplifted crust.

Size, shape: many tens to hundreds of kilometres across; irregular in shape.

Sediment fill: very thick and highly varied, with strong lateral facies changes; usually fluvial at the margins, commonly passing into deep-marine sediment gravity-flow deposits; sometimes the basin becomes sealed off from the ocean, so that facies include lacustrine sediments.

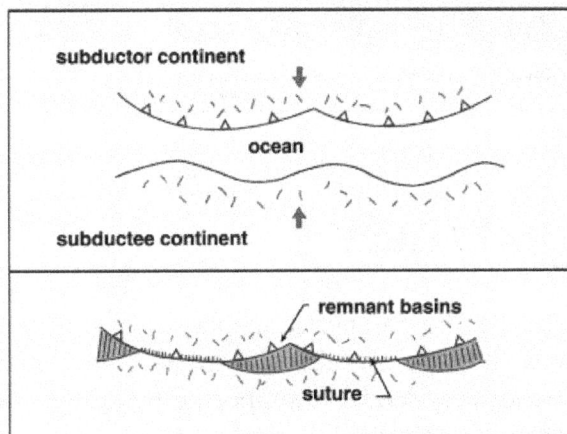

## Pull-Apart Basins

Location and tectonic setting: Locally along major strike-slip or transform faults, either in continental crust or in oceanic crust.

Processes: If a strike-slip fault is stepped or curved rather than straight, movement along it tends to produce tension, where the sense of the curvature and movement are such that the walls of the fault are pulled apart from one another (this kind of regime is described as transtensile), or compression, where the sense of the curvature and movement are such that the walls are pushed against one another (this kind of regime is described as transpressive). In the tensional segments, gaps or basins are produced which are filled with sediment from adjacent high crust.

Size, shape: There is a strong tendency for pull-apart basins to be rhomboidal. They range from approximately equidimensional early in their history to elongated later. Widths are kilometres to a few tens of kilometres, and lengths are kilometres to many tens of kilometres. Some basins are even smaller than this.

Sediment fill: The continental crust basins, which are the most significant sedimentologically, are filled by thick non-marine to marine coarse to fine clastics, often as alluvial fans passing into lake deposits or into deposits of restricted marine environments. In some cases thick marine turbidites fill the distal parts of the basin. There is usually sharp variation in facies laterally, and the thickness of the lithologic units may be not much greater than the lateral extent, or even less. Deposition is concurrent with elongation of the basin, so be wary of total section thickness computed by bed-by-bed measurements of the section.

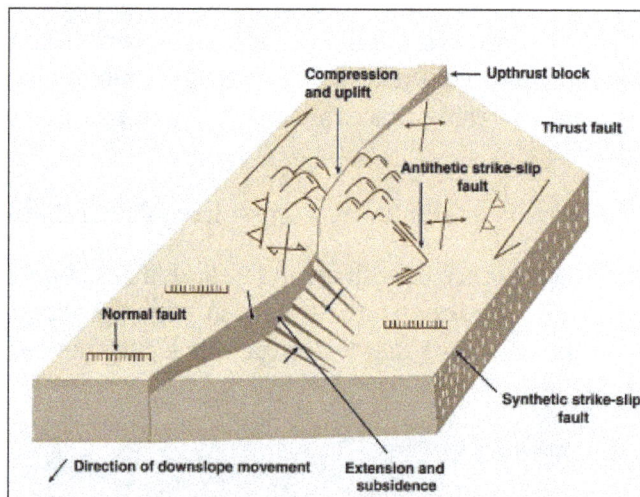

# SEDIMENTARY ISOSTASY

Isostasy is a model that describes how upper lithospheric buoyancy adjusts to any significant change in overburden pressure; it responds to significant additional masses of ice or sediment by subsiding. This equilibrium buoyancy model is supported by a hypothetical mobile magma that exists at depths not yet penetrated by the drilling rig.

Sedimentary isostasy is a related model that addresses how space is created below base level to accommodate additional influxes of sediment. This model assumes that relative sea level remained stable for extensive periods of geological time, despite the continental crust's level changing intermittently throughout the Phanerozoic. The significance of the model is that it can help explain complex geological events observed in sedimentary basins across the globe. In basin analysis it provides a viable alternative to the Sequence stratigraphy model, which implies that sea levels frequently rose above present sea-level and that the crust independently oscillated up and down throughout geological time. It also provides firm theoretical support for the emerging precepts of tectonostratigraphy.

The origin of sedimentary isostasy.

The origin of the sedimentary isostasy model is secured in a letter from John Herschel to Charles Lyell dated 20 February 1836. Herschel was writing from Cape Town, where he was engaged in an astronomical survey of the southern skies. The relationship between crustal uplift and submergence using the limited technical vocabulary of his time. The accompanying sketch shows his firm grasp of a viable mechanism to explain the balance between sediment deposition and adjacent tectonic uplift. Unfortunately his insight only achieved publication two years later in Paragraph I of an Appendix to Charles Babbage's broadly philosophical Ninth Bridgewater Treatise Herschel was a polymath; his fundamental contribution to geology was smothered by his immense contributions in other fields and shortly was forgotten in the heat of geological discovery during that decade.

An early geological cross-section of the Earth's crust.

Charles Babbage and Charles Lyell had already been considering possible relationships between granite and stratified rocks. Babbage applied his Analytical engine to the problem of granite compressibility. In 1837 Lyell inserted a frontispiece in the 5th edition of his *Principles of Geology* that marks a major advance in deep crustal interpretation. Following Herschel's model, it shows a layer at no great depth, where the combination of heat and pressure renders rocks pliable; at greater depths still, they become completely degraded to a mobile granite melt. The lateral flow of melt implied by that diagram provided compelling support for Herschel's concept of sedimentary isostasy.

Homeward bound from his voyage around the world aboard HMS Beagle 1831–1836, Charles Darwin HMS Beagle called at Cape Town on 3 June 1836 and visited Herschel accompanied by Captain FitzRoy. On return to England, Darwin reported some key evidence observed along the shores of Patagonia

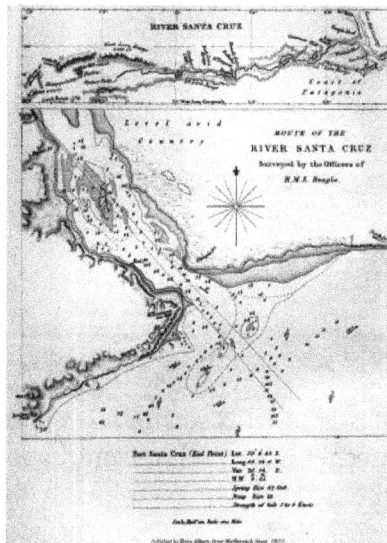

Chart showing part of the Patagonia coast mapped by HMS Beagle in the 1830s.

Considering the enormous power of storm waves, it is unlikely that a sedimentary deposit could have survived the ordeal of beach erosion, unless before uplift, it had considerable thickness and a wide extent offshore. If it were uplifted slowly, the coast should have faced the present ocean with a gentle slope, rather than by a series of steps. This realisation led Darwin to expand the description of sedimentary isostasy: 'Also, on a moderately shallow bottom, where most living marine creatures are found, it seems impossible that a thick expanse of sediment could be deposited widely, unless its bottom sank sufficiently to accommodate successive layers. I am inclined to believe these subsiding movements were accompanied elsewhere by those of elevation'.

Amanz Gressly (1814–1865) studied strata in the Swiss Jura Mountains and described a vertical succession of *groups* or *terrains* there. He then traced beds in horizontal dimensions and encountered marked lateral changes in lithology and paleontology in each stratum. Addressing these lateral changes, he proposed (1838) the term *facies* and identified sandy, muddy, littoral pelagic and other facies, describing them in terms of both sedimentary type and their inferred environment of deposition.

The next significant step was made by the Russian geologist Nikolai A. Golovkinsky (1834–1897). He formulated his *Reverse* rule, which states that the boundaries of lithologically and palaeontologically similar layers regularly *slide* in time when traced laterally. This rule substantiated the primary asynchrony of implied biostratigraphic boundaries. Thus, age-related disposition of the boundaries was associated not only with facies differences in same-aged layers, but also with the palaeobiogeographic features of their faunas. Golovkinsky realized that this contradicted "conventional" views about the process of sedimentation, concluding:

The common belief in the sequence of formation of successive layers 'is not true.'

In 1894 Johannes Walther enunciated the *Law of Correlation of Facies*, which in essence states that within a single sedimentary cycle, the same succession of facies observed in a vertical succession is also present laterally. This fundamental principle traces its origin from the earlier studies of Gressley and Golovkinsky, and *Walther's Law* remains a guiding principle of European stratigraphy.

In 1905 Amadeus William Grabeau|Grabeau described the key concept of disconformity and ideas of stratigraphic onlap and offlap. He was much influenced by the works of Walther and he describes the outcome of Walther's Law in a subsiding sedimentary basin: it will be seen that no two portions of the bed along a line transverse to the seashore will be of the same age, each seaward portion will be younger than that lying next to it nearer the land. Thus the formation line, limiting the basal conglomerate or sandstone, will run diagonally upward through the planes of synchronous deposition. However, when illustrating a marine transgression event he failed to appreciate that Walther's Law would not apply if sea level were rising slowly to form an inverted facies sequence:

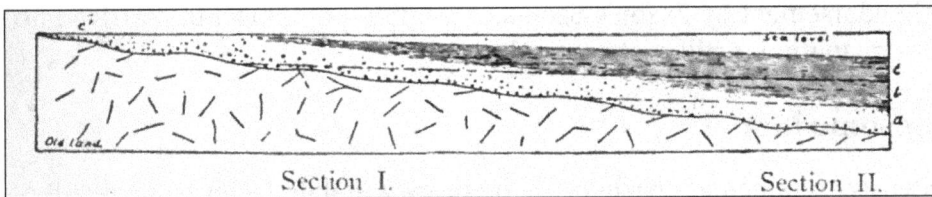

Regular marine progressive overlap on an old land surface illustrates Grabau's ideas of transgression.

land-ward migration of the coarser deposits under uniform conditions (implies) that the changes in any given bed, from the shore sea-ward, will be exactly duplicated by the changes in a given vertical section from the base upward the coarse bed deposited directly upon the old sea-floor of crystalline rock is succeeded upward by a somewhat finer bed, since coarse material has, by continuous subsidence, migrated further landward. Grabeau's transgressive model cannot form a valid basis for cyclic sequences because each subsequent crustal uplift above sea level would cause erosion of the previous layer.

Grabeau's regressive model, by contrast, anticipates many characteristics of the Chronosome model:

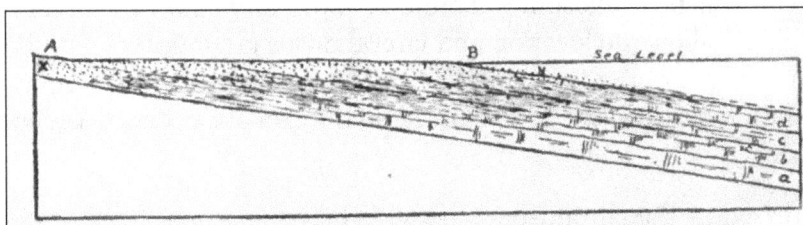

Regressive off-lap and the formation of a sandstone of emergence shows Grabeau's ideas of regression.

For a stationary sea-level, regression of the seashore will take place the shore zone would creep out over the deeper water deposits, the transition from one to the other being rather more abrupt than in the case of a slowly subsiding sea-floor. These relationships are shown in figure, where it will be noted that the diagonal rise of the shore-formed stratum is from the shore sea-ward, whereas in transgressive movements the shore-formed stratum or basal bed rises diagonally land-ward. Such effects are seen in basins (of the continental interior), where the upper beds extend farther toward the mountains from which the material has been derived.

Advances in stratigraphic understanding stagnated after the 1914-18 World War, whilst debate continued about the relative roles of sea level and crustal subsidence. Earth scientists turned to Milankovich theory, meteorite impacts and other catastrophism theories to explain periodic events. During that fallow period for intra-continental stratigraphy, the role of isostasy in crustal movements received a boost from studies of the elevation of Scandinavia during the Holocene, which then was thought to be a response to melting of a thick Pleistocene ice sheet. The removal of that mass postulated by Swedish geologists gave fresh impetus to the theory of isostasy, but any complementary sedimentary loading of North Sea basement was not considered by their investigations.

In 1917, Joseph Barrell was one of the first to discuss composite rhythms of deposition within the context of what is preserved in the stratigraphic record. His concept of a continuously variable base level laid precarious foundations for analytical and seismic stratigraphy, and this approach won eventual endorsement by Exxon's Sequence stratigraphy model that arrived 60 years later, although that explanation is still controversial.

## The Chronosome Model

Thick and extensive sedimentary basins develop below a stable sea level as a result of the deposition and accretion of a succession of sedimentary Chronosomes. A Chronosome is a sedimentary rock unit bounded by planes, each defined by the time of simultaneous deposition of various lithologies. Schultz proposed the term Chronosome for the 3D space bounded by time planes whilst acknowledging that that seismic reflection horizons do not correspond with time planes. Griffiths and Nordlund recognized the value of the chronosome term but considered that lithosomes had greater relevance to seismic correlation, since reflection bounded units are much thicker than the typical chronosome. They considered a chronosome (upper) surface "the nearest we can approach to a paleotopography. The time duration of a chronosome will vary according to the resolution of the observational technique, but at the limit of resolution the chronosome will always provide the optimum predictive unit." The concept had its origins in 1958 following the publication of Time Stratigraphy by Harry E Wheeler  Wheeler diagrams are cross-sections that show the geographic location and order in which chronosomes, bounded above and below by discontinuities were deposited. Wheeler charts show the location and lateral facies gradation of strata and the sequence of deposition in an arbitrary brief time interval. Current examples do not depict the thickness or vertical facies gradation between successive units, so they do not enhance understanding of cyclic deposition.

The problem addressed by the chronosome model is how to account for repeated groups of strata, each group having a dominant component containing fossils of the upper neritic zone that imply marine recession, yet collectively the cyclic sequence has accumulated to vast thicknesses well below sea level. So, whilst sea level appears to be retreating, paradoxically and periodically it returns to exactly the same base level.

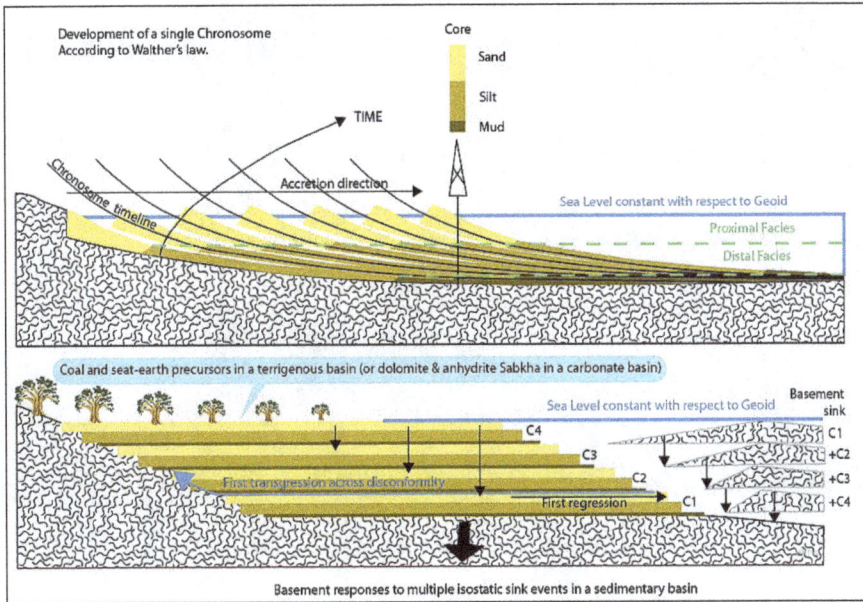

Single chronosome development and cyclical chronosomes in subsiding basin.

Then in a quasi-organic process, it reproduces the same facies sequence with a similar thickness to its predecessor.

In an outcrop or well boring, a bed of sand resting on silt and mudstone taken together forms a distinctive clastic chronosome sequence. Each vertical sequence of facies is closely related to the downslope lateral sequence of facies. Thus, at one instant in time, sands are deposited close to the shore, silts and shales form the seabed farther offshore, illustrating the multidimensional facies relationship first recognized in the Jura Mountains by Gressley, 1838. Over millennia, as sand facies accumulates seawards, previously distal facies get buried and the coastline advances on top of the silt and mudstone. This relationship between vertical and horizontal facies follows Walther's Law. Surfaces that portray levels of simultaneous sediment deposition curve upwards from the horizontal, where vertical accretion is slow in deeper water and become steeper in shallow water where the beach front progrades towards offshore. A similar process operates in carbonate basins, where sabkha facies above the shoreline is followed seawards by bioclastic carbonate sand in shallow water, grading to a lime mud facies farther offshore. Here the driver of the system is the growth and accumulation of a bioclastic facies that grows organically, accumulates in a narrow coastal beach, and gradually extends offshore across the shelf until the single chronosome addition to basin mass exceeds a trigger point that initiates an apparently rapid transgression.

Episodes of subsidence allow the entire basin to accommodate each repetition of chronosome deposition; these explain the accumulation space paradox outlined above. Thus, after the crust subsides rapidly by say 50 to 100 metres, marine water floods across the most recent chronosome, a new coastline is established farther up the sloping craton and seaward regression begins again.

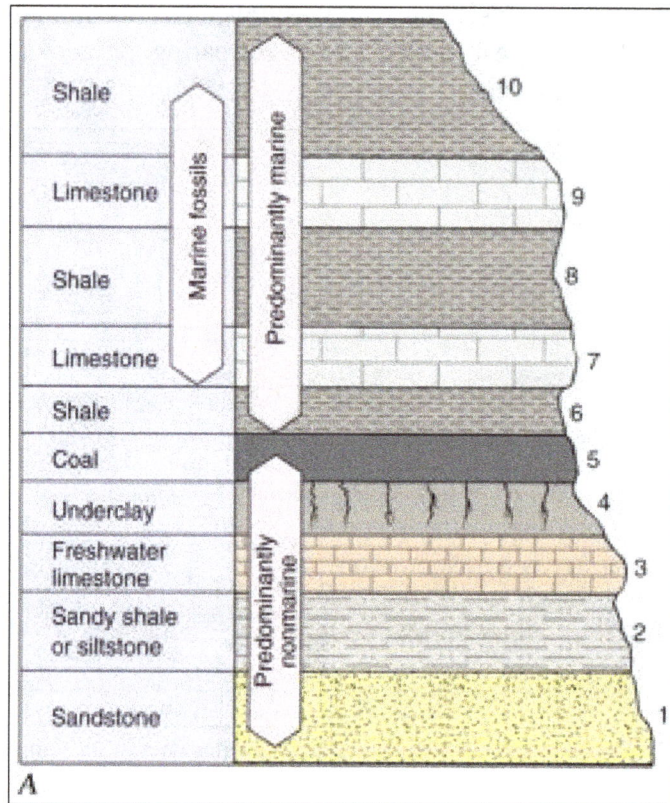

The ideal cyclothem.

On the deeper shelf each chronosome would comprise thinner distal facies and the distal seabed slope would become proportionately steeper.

## Cyclothems or Chronosomes

The cyclothem term is an inexact equivalent of a chronosome model that forms a single regressive sequence. However the cyclothem term has led to much confusion through mis-application in Sequence stratigraphy. It often leads to consistent though invalid boundary planes in its definition of a regressive cycle and perhaps should be regarded as deprecated or obsolete. A thorough historical review of the role of cyclothems was made in 1964 by Marvin Weller for Kansas Geological Survey: This makes it clear that boundaries chosen to define cyclothems were often inconsistent *ad hoc* choices that lack the constraints of an overall depositional, geoid and diastrophic model.

Stratigraphic boundaries should respect both biostratigraphy and facies analysis, which require detailed regional knowledge. This is beyond the scope of seismic stratigraphy and remains difficult even in cored wells. Consequently, notional cyclothem transgressive boundaries have often been chosen at the wrong 'turning point' in a repeating sequence of facies. In addition, unsubstantiated attempts have been made to explain cyclothems by reference to Milankovich cycles or repetitive climate changes. In recent years many stratigraphers prefer an Allostratigraphic model approach, which remains receptive to possible changes in relative sea level. Allostratigraphy is based on the sub-surface correlation of chronstratigraphically significant surfaces through rocks of varying lithologies. It is a way of defining and naming discontinuity bounded rock successions, emphasising unit mappability rather than origin in the context of sea level change. At present, a quasi-uniformitarian approach to stratigraphy using the chronosome model is preferable, pending better definition of geoid changes.

## Correlation across a Stable Basin

In a relatively stable carbonate basin such as the Persian Gulf, scores of stacked chronosomes can be correlated, often bed by bed across the region. Their ages range from the Jurassic Arab Dharb Formation through to the Miocene. Successive chronosomes stack up to four kilometres in thickness, becoming thinner over rising salt domes. Regionally and collectively the Mesozoic and Tertiary succession thins towards the Arabian Craton and thickens towards Iraq and the Zagros Mountains. Repeated small basin floor subsidences provide accommodation space for the initiation and spread of their respective chronosome increments. In areas of more complex structure, individual chronosomes can be correlated within specific fault blocks and otherwise can assist restoration of folds and faults to their depositional forms. A significant characteristic of the Persian Gulf sequence is the prolific development of cyclic sedimentation. The causes of widespread cyclothem occurrence are still disputed by stratigraphers but the implications of the Sedimentary Isostasy model may lead to a better understanding of this issue.

## Chronosome Sea Level Model

There is no stratigraphic or geomorphological evidence that provides unequivocal support to a sea level significantly higher than the current maximum tidal range. Any apparent exceptions to this rule can be explained otherwise by epeirogenic or orogenic crustal uplifts or by local gravity anomalies that have disturbed the present geoid model. Chronosome theory requires that sea level was fixed relative to a geoid with an equatorial diameter that remained stable and constant with only occasional catastrophic interruptions during the Phanerozoic Era. The chronosome model acknowledges that the Earth's geoid was disturbed intermittently by plate tectonics and meteorite impacts throughout that period. Events such as the global Maastrichtian Transgression established a new and stable base level that defined a new geoid based on a revised Earth's equatorial diameter. Tectonic activity related to such catastrophic disturbances caused disconformities that interrupt the chronosome model, which resumed its cyclical progress when the geoid attained equilibrium.

## Chronosome Isostasy Model

At a certain point in time the basin-wide thickness and mass of a single chronosome exceeds a nominal isostatic threshold; the entire basin sinks simultaneously, sea water sweeps across the previously deposited chronosome sand layer and a new coastline is established farther up the craton slope, causing the recessive deposition process to begin again. This isostatic model assumes that mass per unit area is evenly distributed across most of the basin because the heavier and thicker proximal facies gradually spreads across most of the basin. However Bailey Willis pointed out that: *Each region has experienced an individual history of diastrophism, in which the law of periodicity is expressed in cycles of movement and quiescence peculiar to that region.* It follows that global correlation using Sequence Stratigraphy may not be reliable, except for very broad intervals of time.

# Cyclicity Mechanism Underpinning

The structure of the Viking Graben and adjacent areas was established from seismic and well data during the late 20th century.

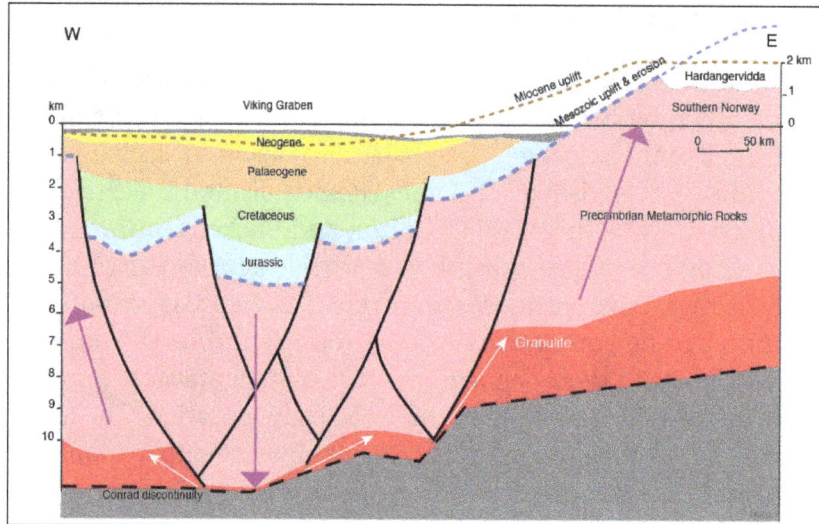

Subsidence in the Viking Graben balanced by uplift in Norway 2019.

Its relationship with the Scandian high level peneplain was established only in 2018 when Japsen et al. described evidence on the Hardangervidda Mountains in southern Norway that helped validate Herschel's model.

Herschel's letter apparently influenced Charles Lyell's Principles of Geology Book frontispiece. This shows rocktype sequences in a stable sedimentary basin. Below about 3,000 metres of porous water- and hydrocarbon-saturated sedimentary rocks are a further 2,000 metres of sediment with porosity mostly filled by diagenetic cement. Beyond that level, sediments are metamorphosed by temperature and pressure to form quartzites, schists and marbles with ground water absorbed as crystal components. Deeper still, rocks are encountered that have decomposed into a migmatite, otherwise a melange of metamorphic rock traces in a granitic matrix. Finally, the combination of temperature and pressure exceeds the phase transition threshold of supercritical water, which assists in the production of an extremely mobile granitic melt at a depth of 10 to 15 kilometres. The current chronosome model assumes that the thickness of this fluid zone is limited by a very precise combination of physical and chemical factors to form a thin layer, implied by the amount of subsidence to be less than 100m. thick. This zone likely coincides with the Conrad discontinuity. Above that level, any reduction in temperature and pressure causes water to drop below its supercritical phase threshold and minerals in the granite melt crystallize in sequence. This occurs not far above the Conrad level because granite melt rarely reaches the surface as rhyolite lava; if it should reach the surface, it can cause a catastrophic eruption. At the Conrad Discontinuity granite magma encounters a process of anatexis or differentiation, by which is meant gravity separation of the basic heavy minerals, which sink below the residual granite magma to form a restite of immiscible basic magma. The separate granulite residue has its mobility further enhanced as a consequence, and overburden pressure causes it to seek a lateral passage away from the centre of the basin. As the granite magma's initial volume is expelled laterally, the entire basin sinks into the space vacated by magma, thus placing the next level of transitional migmatite directly on top of the restite, where it begins to melt and perpetuate cyclical processes throughout the basin. There is usually an ancient craton on at least one side of an extensive sedimentary basin. When the lateral impetus of granite magma reaches a craton boundary it splits the craton horizontally at Conrad level and raises or tilts a fault block upwards to form a new horst feature. Successive magma injections beneath it raise the horst block further by stages and related uplift and erosion of the block contribute a further influx of sediment to form a new chronosome in the basin.

## Practical Applications of the Chronosome Model

Chronosomes are the most productive environments for the growth of plant and animal life and formed source rocks for petroleum in both terrigenous and carbonate buried sediments. Extensive plains that extend behind the coastline were a favourable environment for prolific plant life that became buried quickly by each marine transgression. In the Upper Pennsylvanian, Permian and Cretaceous coal and lignite were sources for waxes and resins that produce asphalt and methane under temperature and pressure conditions of deep burial. Rivers carried decaying vegetable humus into an adjacent basin where it became buried in association with silts and muds. The intertidal zone is a prolific habitat for seashells, crustaceans, bird droppings and sea grass. Below wave base on temperate, terrigenous shelves, kelp forests restrict the offshore flow of sediments. This increases the rate of sediment deposition and their buried remains further contribute to waxy source rocks. Blue and green marine algae, diatoms and lithothamnium also offer promising source material for petroleum in both temperate terrigenous and tropical carbonate basins. On carbonate platforms the coastal plain is a predominantly sterile sabkha environment. Reef-building corals seashells and crustaceans accumulate on its shallower neritic zones, whilst foraminifera, sponges and algae get buried beneath a rain of carbonate mud in deeper water. Clearly, the productivity of organic life greatly assists gravity, insolation and climate in advancing the accumulation of successive chronosomes throughout the world.

Fluid flow in a petroleum reservoir is influenced by the distribution of porosity and permeability in its component strata. The identification of 'flow units' within this sequence can help determine appropriate layers for use in reservoir modelling. These models are based on a three-dimensional grid of cells. The flow characteristics and fluid contents of these cells determine the outcome of a computer program that simulates the effects of reducing and increasing pressure in various parts of the reservoir. This reservoir model is based on modelled geometry of facies, established from cores in early development wells. Thus its stratigraphic parameters are based on layers that combine a Chronosome model with a Wheeler Diagram. Practical limitations on cell size ensures that initial models provide only a provisional, yet indispensable likeness of facies distribution.

# SEDIMENTARY BASIN ANALYSIS

Sedimentary basin analysis is a geologic method by which the history of a sedimentary basin is revealed, by analyzing the sediment fill itself. Aspects of the sediment, namely its composition, primary structures, and internal architecture, can be synthesized into a history of the basin fill. Such a synthesis can reveal how the basin formed, how the sediment fill was transported or precipitated, and reveal sources of the sediment fill. From such syntheses models can be developed to explain broad basin formation mechanisms. Examples of such basinal environments include backarc, forearc, passive margin, epicontinental, and extensional basins.

Sedimentary basin analysis is largely conducted by two types of geologists who have slightly different goals and approaches. The petroleum geologist, whose ultimate goal is to determine the possible presence and extent of hydrocarbons and hydrocarbon-bearing rocks in a basin, and the academic geologist, who may be concerned with any or all facets of a basin's evolution. Petroleum industry basin analysis is often conducted on subterranean basins through the use of reflection seismology and data from well logging. Academic geologists study subterranean basins as well as those basins which have been exhumed and dissected by subsequent tectonic events. Thus academics sometimes use petroleum industry techniques, but in many cases they are able to study rocks at the surface. Techniques used to study surficial sedimentary rocks include: measuring stratigraphic sections, identifying sedimentary depositional environments and constructing a geologic map.

An important tool in sedimentary basin analysis is sequence stratigraphy, in which various sedimentary sequences are related to pervasive changes in sea level and sediment supply.

# PULL-APART BASIN

Pull-apart basins are structural depressions that localized on the geometric irregularities along strike-slip faults where the master fault overstepping or bending.

Strike-slip faults form linear and continuous fault systems, but they are typically segmented, resulting in localized regime changes across a variety of discontinuities or steps. Pull-apart basins have been recognized originally along major strike-slip faults throughout the world where the segmented faults cause extension/transtension. Quennell, who was the first to recognize a pull-apart basin without using this term, proposed that the Dead Sea is a void in the crust caused by the overlapping segments of the Dead Sea fault zone. For the same purpose, the first use of the term "pull-apart basin" by Burchfiel and Stewart was chosen for the interpretation of the Death Valley basin. The term pull-apart basin is a well-used and well-understood term for strike-slip basins both in marine and terrestrial environments.

## Model for Pull-Apart Basin Development

As indicated above, there are several synonyms of pull-apart basins that reproduced from the identified tectonic, structural, geometric, and geomorphic features of pull-apart basins in nature. Tectonically, the basins are located along strike-slip faults and transform zones; structurally, they form when strike-slip faults bend or overstep; geometrically, while the overstepping master strike-slip faults are parallel or subparallel, the secondary basin-bounding faults perpendicular or diagonal to the main faults give the basin its characteristic shape; and geomorphologically, they are sharp-bounded and deep depressions. All these characteristics of pull-apart basins in nature help us to draw an ideal model for pull-apart basin that represents its mechanism of formation.

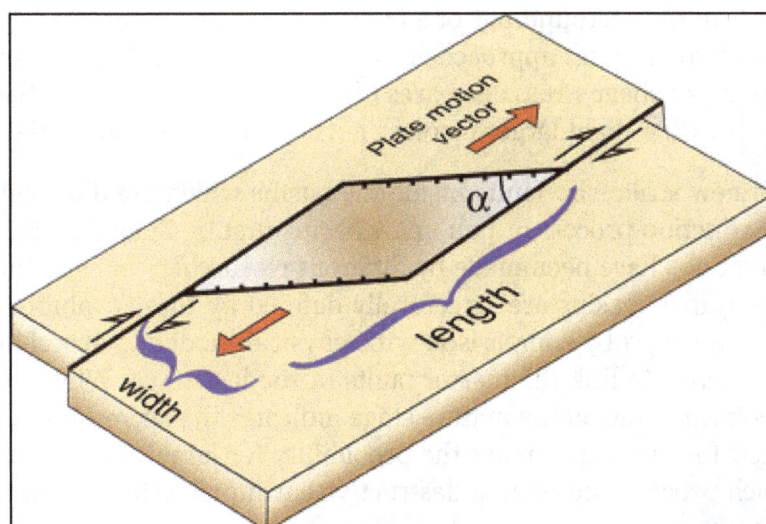

Formation and geometric model of pull-apart basin.

Detailed properties of natural pull-apart basins have been obtained through using experimental models. Although experimental laboratory studies have been used to investigate strike-slip faults since approximately the first quarter of twentieth century, they have been used to simulate the mechanism and geometries of pull-apart basin formation in the last quarter. Because these basins differ considerably from simple strike-slip mechanism, their development processes share properties with both strike-slip and extensional settings. Physical and/or numerical model studies have been a profitable approach, producing good geometric matches with natural examples of pull-apart basins. While the physical (analogue) modeling experiments used a variety of apparatus and different materials (e.g., sand, clay, and rock), the numerical experiments use a wide range of parameters and quantitative values. Thus, the results are helpful to understand the mechanical and geometrical aspects of natural large-scaled cases under different conditions.

## Mechanical Aspects

As detailed above, the basic mechanism of pull-apart basin development can be defined simply as local extension between two overstepping strike-slip faults. Distributed strike-slip and Riedel shear mechanisms are two alternative mechanisms that have been proposed for pull-apart basin formation. Distributed shear deformation is developed in weak layers, such as evaporates or shales, above a broad strike-slip fault zone in the basement, and pull-apart basins can develop at releasing steps during fault interaction, coalescence, and linkage. The Riedel shear mechanism causes development of pull-apart basins along Riedel faults connected by segments of strike-slip faults in the basement.

Independent of the formation mechanism, the extension which develops within pull-apart basins depends on the local crustal rheology. While the displacement along the master strike-slip faults links to each other by normal faults in a brittle crust, in a ductile crust it accommodates by other mechanisms (e.g., continuous extension inside the basin; Reches). Another necessary condition for a pull-apart basin formation is a mechanical detachment between the brittle part of the crust and rheologically strong uppermost mantle. The strength of décollement zones of pull-apart basins, which are observed or postulated to form within ductile or viscous zones at depth, also affects structural development in most tectonic regimes and likely exhibits some control on the development process.

The elastic stress field in the surroundings of a fault step was firstly analyzed by Rodgers and Segall and Pollard with two different approaches. However, their conclusions were consistent with each other: while the mean shear stress decreases inside the pull-apart basin, the maximum stress increases at fault ends with a slight larger increase outside the basin boundaries.

The development of new strike-slip faults inside the basins which are diagonal to master faults, has defined as the extinction process of pull-apart basins first by Zhang et al. Similar tendencies and real-time observations have been made by different researchers in experimental models and earthquakes. The pull-apart basins are structurally defined by three evolutionary stages: incipient, early, and mature stages by comparison with physical modeling experiments. Dramatically this type of basins tends to link the master faults in the last stage. While the development of through-going cross-basin faults in the mature stage indicates the decreasing activity along major boundary faults, their formations increase the probability for large-magnitude earthquakes. The best examples of such process and related destructive earthquakes have been seen in China and Turkey in the last century.

## Geometrical Aspects

The shape, fault system, and sedimentation features of a pull-apart basin depend upon the geometry associated with the step in the master strike-slip fault system. Likewise, the geometry of a pull-apart basin is controlled by the amount of master fault overlap, separation, and displacement, and the size of the basin is varied and well defined by these parameters. Therefore to understand their anatomies and evolutions, we need to define some geometric parameters because many uncertainties complicate their interpretations. shape, scale, and angular characteristics of pull-apart basins are useful to discuss their initiation and structural.

Pull-apart basin structures are usually rhomb-shaped depressions, but they vary from the lazy Z- or S-shaped to spindle- or almond-shaped to extreme-shaped examples in nature proposed that the rhomboidal form of pull-apart basins results from the lengthening of an S- or Z-shaped basin with increased master fault overlap. Regardless of offset geometry, they evolve progressively from narrow grabens bounded by oblique-slip-linked faults to wider rhombic basins. Within these shapes, they can be asymmetric, symmetric, or hybrid forms depending on the activity and connectivity of basin-bounding faults.

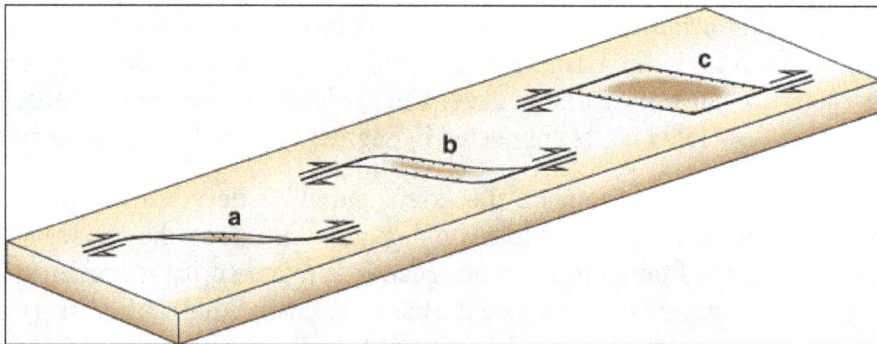

Typical shapes of pull-apart basins: (a) spindle-shaped, (b) lazy Z-shaped, (c) rhomboidal.

Within angular meaning, the results of substantial experimental studies and natural pull-apart basins from all over the world represent the acute angles ($\alpha$ angle in figure) between the oblique bounding faults and the master strike-slip faults of pull-apart basins are generally clustered at 30°–35°. Otherwise, the boundary condition (i.e., pure strike-slip, transtension, transpression) is the major parameter determining the acute angle thereby the shape of the basin.

Pull-apart basins have a specific scale ratio. Aydın and Nur showed a well-defined linear correlation between the length and width of pull-apart basins with a value of 3:1 through analyzing of over 60 natural examples and proposed that the value may vary widely, depending on the structural, physiographic, or active dimensions of the basin. In addition, experimental models show similar results with an approximately constant length versus width ratio. In the same vein, a step further, a three-dimensional relationship between the length, width, and depth of pull-apart basins has suggested by Gürbüz) through using the natural examples.

Pull-apart basins represent the same mechanical and geometrical features both in marine and terrestrial environments and sometimes they represent a complex basin system scene. Marmara Sea of Turkey is located on the North Anatolian fault zone which is one of the world's famous strike-slip faults with its length of ~1,200 km and its internal structure includes such a basin system. Each of the basins has developed by the strike-slip tectonics of the North Anatolian fault zone and represents similar geometric features while varying in shapes, sizes, and also depths.

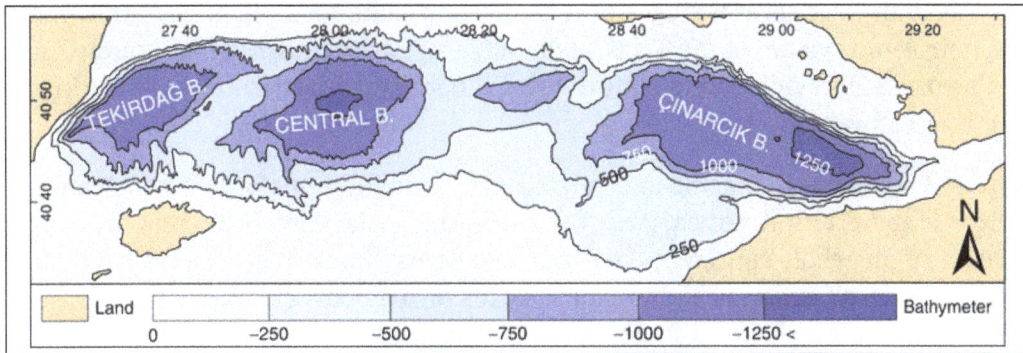

Bathymetric map of the Marmara Sea pull-apart basin system.

## Sedimentary Aspects

Sedimentation features of pull-apart basins are generally related to kinematic, geometric, geomorphic, and crustal properties. Strike-slip faults that bound these basins have a large component of normal slip. Thus, while the basin inside is represented by a deep depression, the basin margins surrounded by steeper uplifted and deeply eroded highlands.

In continental pull-apart basins, along the strike-slip margins of the basin, coarser sediments in the form of talus breccia and debris-flow-dominated alluvial fans grade into fine-grained sediments toward the center of the basin. As a result of such geomorphic and geometric characteristics, higher sedimentation rates, greater thicknesses in smaller areas, displaced fan-source relationships and skewed fans, and depocenter migrations differentiate pull-apart basins from other extensional depressions. In marine environments, similar sedimentary patterns are valid for pull-apart basins by coarser sediments along basin margins transported through submarine canyons instead of rivers and submarine landslides along fault scarps and fine sediments toward the deepest parts Basins far away from efficient sediment sources may include pelagic sediments and deposits of gravity mass movements.

# SEDIMENTARY BASINS AND PETROLEUM SYSTEMS

The modern definition and classification of sedimentary basins and associated petroleum systems is principally based on concepts of plate tectonics. The plate-tectonic cycle, in its initial divergent setting, begins typically with intracratonic rifting, followed by opening of oceans and development of passive continental margins. Subduction, continental collision, and orogenesis mark the second, convergent part of the cycle. During the course of the plate-tectonic cycle, various sedimentary basins and related petroleum systems evolve in this dynamic global system. The hydrocarbon potential of each principal basin is determined by its tectonic and depositional history, which relates to succession of critical stages of the plate tectonic cycle: rifting, sagging, drifting, subduction, and collision. Among many types of basins, the rift and intracratonic rift-sag basins, and passive continental margins of the divergent settings, the foreland basins and fold and thrust belts of the convergent settings, and some borderland basins of the transform margins represent the most petroliferous provinces of the world. The petroleum systems, including all processes from source-rocks deposition to final entrapment of hydrocarbons, evolved in a single stage of a plate tectonic cycle; e.g., in simple rift basins or during two or more stages of a cycle, e.g., on passive continental margins with stages of rifting, sagging, and drifting. In some cases, the formation of a petroleum system even extends over several stages of two different plate tectonic cycles. An example of the latter is the petroleum system of the Trias basin in North Africa, in which the source rocks deposited in the Paleozoic-Hercynian cycle matured and generated hydrocarbons only during the Mesozoic to Cenozoic Tethyan/Alpine cycle.

Various types of sedimentary basins and related petroleum systems are demonstrated on examples from North and South America, Africa, Europe, and Asia. Also presented are maps of sedimentary basins of the whole continents of Africa and South America as examples of application of the plate-tectonic classification of sedimentary basins on a global scale.

Definition of sedimentary basins and petroleum systems in terms of plate tectonics not only enables better categorization and integration of data, but also serves as a predictive tool in evaluation of various properties and in search for new potential hydrocarbon plays. Well-defined hydrocarbon habitats of certain plate-tectonic settings will serve as potential models for other, less known basins and exploration plays. However, it is not the categorization and classification of sedimentary provinces and petroleum systems itself but the process of analyzing data within a universal concept of global tectonics that enhances creative thinking and consequently improves the chances for successful exploration.

## References

- Sedimentary-basins, geologylearn.blogspot.com: Retrieved 4 April, 2019

- Quayyum, F.C.; Catuneau, O. (2017). "The Wheeler diagram, flattening theory, and time". Journal of Marine and Petroleum Geology. 86: 1417–22851430

- How-are-sedimentary-basins-formed: geologylearn.blogspot.com, Retrieved 5 May, 2019

- Quayyum, F.C.; Catuneau, O. (2017). "The Wheeler diagram, flattening theory, and time". Journal of Marine and Petroleum Geology. 86: 1417–22851430

- Classification-of-sedimentary-basins: geologylearn.blogspot.com, Retrieved 6 June, 2019

# 6

# Uses of Sediments

There are various uses of sediments. They play an important role in the formation of soil which is essential for growing crops. Sedimentary rocks are used in civil engineering for building construction, cement production, pavement and road production, etc. The topics elaborated in this chapter will help in gaining a better perspective about these diverse uses of sediments.

## REAL-LIFE APPLICATIONS OF SEDIMENTS

### Sediments and Dust Bowls

Sediment makes possible the formation of soil, which of course is essential for growing crops. Therefore it is a serious matter indeed when wind and other forces of erosion remove sediment, creating dust-bowl conditions. The term "Dust Bowl," with capital letters, refers to the situation that struck the United States Great Plains states during the 1930s, devastating farms and leaving thousands of families without home or livelihood.

During the late 1990s, some environmentalists became concerned that farming practices in the western United States were eroding sediment, putting in place the possibility of a return to the conditions that created the Dust Bowl. However, in August 1999, the respected journal *Science* reported studies showing that sediment in farmlands was not eroding at anything like the rate that had been feared. Soil scientist Stanley Trimble at the University of California, Los Angeles, studied Coon Creek, Wisconsin, and its tributaries, a watershed for which 140 years' worth of erosion data were available. As Trimble discovered, the rate of sediment erosion in the area had dramatically decreased since the 1930s, and was now at 6% of the rate during the Dust Bowl years.

Some studies from the 1970s onward had indicated that farming techniques, designed to improve the crop output from the soil, had created a situation in which sediment was being washed away at alarming rates. However, if such sediment removal were actually taking place, there would have to be some evidence—if nothing more, the sediment that had been washed away would have had to go somewhere. Instead, as Trimble reported," We found that much of the sediment in Coon Creek doesn't move very far, and that it moves in complex ways." The sediment, as he went on to explain, was moving within the Coon Creek basin, but the amount that actually made it to the Mississippi River (which could be counted as true erosion, since it was removing sediment from the area) had stayed essentially the same for the past 140 years.

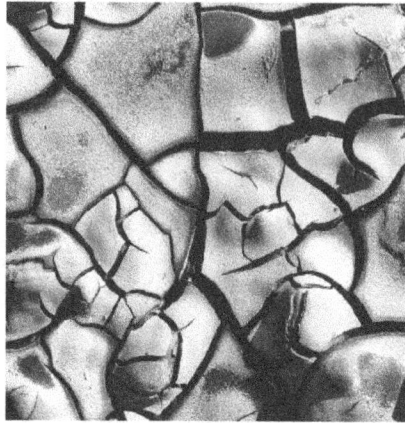

S Edimentary Structures Remaining In A Dried River Bed . C Lay Soils Crack
as they Lose Moisture and Contract when Trapped Water Evaporates as the Result of Drought.

## Deposition and Depositional Environments

Eventually everything in motion—including sediment—comes to rest somewhere. A piece of sediment traveling on a stream of water may stop hundreds of times, but there comes a point when it comes to a complete stop. This process of coming to rest is known as deposition, which may be of two types, mechanical or chemical. The first of these affects clastic and organic sediment, while the second applies (fittingly enough) to chemical sediment.

In mechanical deposition, particles are deposited in order of their relative size, the largest pieces of bed load coming to a stop first. These large pieces are followed by medium-size pieces and so on until both bed load and suspended load have been deposited. If the sediment has come to a full stop, as, for instance, in a stagnant pool of water, even the finest clay suspended in the water eventually will be deposited as well.

Unlike mechanical deposition, chemical deposition is not the result of a decrease in the velocity of the flow; rather, it comes about as a result of chemical precipitation, when a solid particle crystallizes from a fluid medium. This often happens in a saltwater environment, where waters may become overloaded with salt and other minerals. In such a situation, the water is unable to maintain the minerals in a dissolved state (i.e., in solution) and precipitates part of its content in the form of solids.

## Depositional Environments

The matter of sediment deposition in water is particularly important where reservoirs are concerned, since in that case the water is to be used for drinking, cooking, bathing, and other purposes by humans. One of the biggest problems for the maintenance of clean reservoirs is the transport of sediment from agricultural areas, in which the soil is likely to contain pesticides and other chemicals, including the phosphorus found in fertilizer. A number of factors, including precipitation, topography, and land use, affect the rate at which sediment is deposited in reservoirs.

The area in which sediment is deposited is known as its depositional environment, of which there are three basic varieties: terrestrial, marginal marine, and marine. These are, respectively, environments on land (and in landlocked waterways, such as creeks or lakes), along coasts, and in the open ocean. A depositional environment may be a large-scale one, known as a regional environment, or it may be a smaller subenvironment, of which there may be hundreds within a given regional environment.

## Sedimentary Structures

There are many characteristic physical formations, called sedimentary structures, that sediment forms after it has reached a particular depositional environment. These formations include bedding planes and beds, channels, cross-beds, ripples, and mud cracks. A bed is a layer, or stratum, of sediment, and bedding planes are surfaces that separate beds. The bedding plane indicates an interruption in the regular order of deposition.

Channels are simply depressions in a bed that reflect the larger elongated depression made by a river as it flows along its course. Cross-beds are portions of sediment that are at an angle to the beds above and below them, as a result of the action of wind and water currents—for example, in a flowing stream. As for ripples, they are small sandbar-like protuberances that form perpendicular to the direction of water flow. At the beach, if you wade out into the water and look down at your feet, you are likely to see ripples perpendicular to the direction of the waves. Finally, mud cracks are the sedimentary structures that remain when water trapped in a muddy pool evaporates. The clay, formerly at the bottom of the pool, begins to lose its moisture, and as it does, it cracks.

## The Impact of Sediment

It is estimated that the world's rivers carry as much as 24 million tons (21,772,800 metric tons) of sediment to the oceans each year. There is also the sediment carried by wind, glaciers, and gravity. Where is it all going? The answer depends on the type of sediment. Clastic and organic sediment may wind up in a depositional environment and experience compaction and cementation in the process of becoming sedimentary rock.

On the other hand, clastic and organic particles may be buried, but before becoming lithified (turned to rock), they once again may be exposed to wind and other forces of nature, in which case they go through the entire cycle again: weathering, erosion, transport, deposition, and burial. This cycle may repeat many times before the sediment finally winds up in a permanent depositional environment. In the latter case, particles of clastic and organic sediment ultimately may become part of the soil.

A chemical sediment also may become part of the soil, or it may take part in one or more biogeochemical cycles These chemicals may wind up as water in underground reservoirs, as ice at Earth's poles, as gases in the atmosphere, as elements or compounds in living organisms, or as parts of rocks. Indeed, all three types of sediment—clastic, chemical, and organic—are part of what is known as the rock cycle, whereby rocks experience endlessly repeating phases of destruction and renewal.

## Sedimentary Mineral Deposits

Among the most interesting aspects of sediment are the mineral deposits it contains—deposits that may, in the case of placer gold, be of significant value. A placer deposit is a concentration of heavy minerals left behind by the effect of gravity on moving particles, and since gold is the densest of all metals other than uranium (which is even more rare), it is among the most notable of placer deposits.

Of course, the fact that gold is valuable has done little to hurt, and a great deal to help, human fascination with placer gold deposits. Placer gold played a major role from the beginning of the famous California Gold Rush (1848-49), which commenced with discovery of a placer deposit by prospector James Marshall on January 24, 1848, along the American River near the town of Coloma. This discovery not only triggered a vast gold rush, as prospectors came from all over the United States in search of gold, but it also proved a major factor in the settlement of the West. Most of the miners who went to the West failed to make a fortune, of course, but instead they found something much better than gold: a gorgeous, fertile land like few places in the United States—California, a place that today holds every bit as much allure for many Americans as it did in 1848.

Despite the attention it naturally attracts, gold is far from the only placer mineral. Other placer minerals, all with a high specific gravity (density in comparison to that of water), include platinum, magnetite, chromite, native copper, zircon, and various gemstones. Nor are placer minerals found only in streams and other flowing bodies of water; wave action and shore currents can leave behind what are called beach placers. Among the notable beach placers in the world are gold deposits near Nome, Alaska, as well as zircon in Brazil and Australia, and marine gravel near Namaqualand, South Africa, which contains diamond particles.

An entirely different process can result in the formation of evaporites, minerals that include carbonates, gypsum, halites, and magnesium and potassium salts. (These specific mineral types are discussed in Minerals.) Formed when the evaporation of water leaves behind ionic, or electrically charged, chemical compounds, evaporites sometimes undergo physical processes similar to those of clastic sediment. They may even have graded bedding, meaning that the heavier materials fall to the bottom. In addition to their usefulness in industry and commerce (e.g., the use of gypsum in sheetrock for building), physical and chemical aspects of evaporites also provide scientists with considerable information regarding the past climate of an area.

# OIL/FUEL

Natural gas refers collectively to the various gaseous hydrocarbons generated below the Earth's surface and trapped in the pores of sedimentary rocks. Major natural gas varieties include methane, ethane, propane, and butane. These natural gases are commonly, though not invariably, intimately associated with the various liquid hydrocarbons—mainly liquid paraffins, napthenes, and aromatics—that collectively constitute oil.

Hydrocarbons can also exist in a semisolid or solid state such as asphalt, asphaltites, mineral waxes, and pyrobitumens. Bitumens can occur as seepages, impregnations filling the pore space of sediments (e.g., tar sands of the Canadian Rocky Mountains), and in veins or dikes. Asphaltites occur primarily in dikes and veins that cross sedimentary rocks such as gilsonite deposits in the Green River Formation of Utah. These natural bitumens probably form from the loss of volatiles, oxidation, and biological degradation resulting from oil seepage to the surface. Solid hydrocarbons are of interest to geologists as their presence is a good indicator of petroleum below the surface in that region. Also, solid hydrocarbons have commercial value.

The exact process by which oil and natural gas are produced is not precisely known, despite the extensive efforts made to determine the mode of petroleum genesis. Crude oil is thought to form from undecomposed organic matter, principally single-celled floating phytoplankton and zoo-plankton that settle to the bottom of marine basins and are rapidly buried within sequences of mudrock and limestone. Natural gas and oil are generated from such source rocks only after heating and compaction. Typical petroleum formation (maturation) temperatures do not exceed 100 °C, meaning that the depth of burial of source rocks cannot be greater than a few kilometres. After their formation, oil and natural gas migrate from source rocks to reservoir rocks composed of sedimentary rocks largely as a consequence of the lower density of the hydrocarbon fluids and gases. Good reservoir rocks, by implication, must possess high porosity and permeability. A high proportion of open pore spaces enhances the capacity of a reservoir to store the migrating petroleum; the interconnectedness of the pores facilitates the withdrawal of the petroleum once the reservoir rock is penetrated by drill holes.

# COAL

Coals are the most abundant organic-rich sedimentary rock. They consist of undecayed organic matter that either accumulated in place or was transported from elsewhere to the depositional site. The most important organic component in coal is humus. The grade or rank of coal is determined by the percentage of carbon present. The term peat is used for the uncompacted plant matter that accumulates in bogs and brackish swamps. With increasing compaction and carbon content, peat can be transformed into the various kinds of coal: initially brown coal or lignite, then soft or bituminous coal, and finally, with metamorphism, hard or anthracite coal. In the geologic record, coal occurs in beds, called seams, which are blanketlike coal deposits a few centimetres to metres or hundreds of metres thick.

Many coal seams occur within cyclothems, rhythmic successions of sandstone, mudrock, and limestone in which nonmarine units are regularly and systematically overlain by an underclay, the coal seam itself, and then various marine lithologies. The nonmarine units are thought to constitute the floor of ancient forests and swamps developed in low-lying coastal regions; the underclay is a preserved relict of the soil in which the coal-producing vegetation was rooted; and the marine units overlying the coal record the rapid transgression of the sea inland that killed the vegetation by drowning it and preventing its decomposition by rapid burial. The exact mechanism responsible for generating the rapid episodes of marine transgression and regression necessary to generate coal-bearing cyclothems is not definitively known. A combination of episodic upwarping and downwarping of the continental blocks or global (eustatic) changes in sea level or both, coupled with normal changes in the rate of sediment supply that occurs along coasts traversed by major laterally meandering river systems, may have been the cause.

In any case, coal is a rare, though widely distributed, lithology. Extensive coal deposits overall occur mainly in rocks of Devonian age (those from 408 to 360 million years old) and younger because their existence is clearly contingent on the evolution of land plants. Nevertheless, small, scattered coal deposits as old as early Proterozoic have been described. Coal-bearing cyclothem deposits are especially abundant in the middle and late Paleozoic sequences of the Appalachians and central United States and in the Carboniferous of Great Britain, probably because during this time interval global climates were warm and humid and large portions of the continental blocks were low-lying platforms located only slightly above sea level.

# USES OF SEDIMENTARY ROCKS IN CIVIL ENGINEERING

There are many uses of sedimentary rocks in civil engineering. The main applications of this type of rock is provided below:

## Building Construction

### Sandstone

- Sandstone, which is easy to work with, has been broadly utilized in the construction of buildings specifically in areas where large quantity of sandstone is available.

- For example, The Cliffe Castle Museum in England is composed completely of sandstone. In addition to Red Fort building stone in India.

- Marble is also used mainly in building construction.

- Famous marble buildings are Taj Mahal in India, the leaning tower of Pisa in Italy, the Parthenon in Greece.

### Limestone

Limestone has been used in several important building constructions for instance monuments.

Redfort building in India, Sandstone application in building construction.

Cliffe castle museum in England.

## Structural Wall Construction

Both sandstone and limestone are suitable for the construction of structural walls. Nonetheless, sandstone shall be considered carefully because it might be excessively porous and fragile for load bearing structures.

Wall constructed from sandstone.

## Cement Production

- Limestone is the main source material for the production of Portland cement.

- Shale is used as a component in cement production.

Limestone hauled for cement production.

## Concrete Production

Sedimentary rock used as aggregate in concrete production to withstand pressure.

Limestone aggregate for concrete production.

## Pavement and Road Construction

- Sandstone and limestone have been used for the construction of pavement stone and road stone.

- Shale is used as aggregate in road construction.

- Sedimentary rocks are used in highway roadbed.

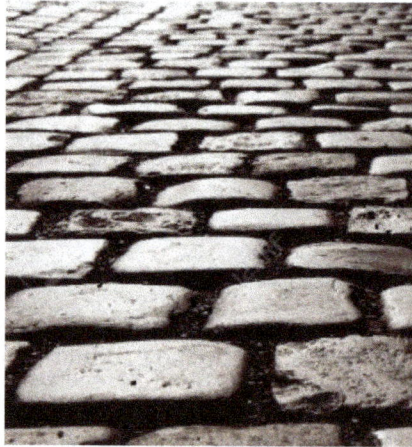

Road stone.

## Tunneling

- Sandstone is suitable type of rock for tunneling especially thick bedded, well cemented, and siliceous or ferruginous sandstone.

- It is strong and easily workable.

- Lining is not needed.

- Sandstone does not influence geological structures and ground condition detrimentally.

- Tunneling is easily progressed in shale formation due to its softness, but proper lining shall be provided.

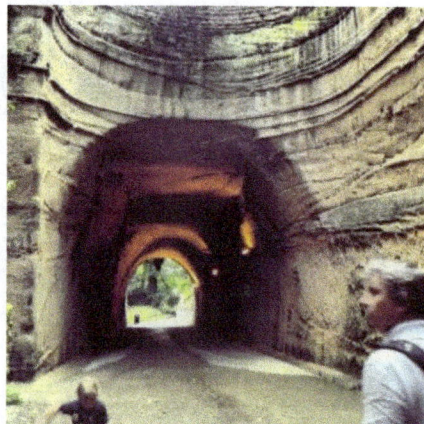

Tunneling work, sedimentary rock.

## Brick and Tile Manufacturing

Shale is used as one components of materials used in the construction of bricks and tiles.

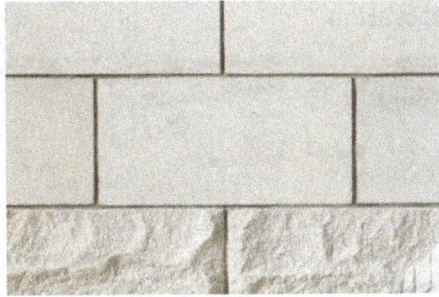

Tile manufactured from limestone.

## Architectural and Monumental Stone

- There are certain types of sedimentary rock that can be used as a architectural and monumental stone for example Portland stone (a white -grey limestone).

- It can withstand weathering affects adequately. More importantly,Portland stone can be cut and craved comfortably by masons which is a crucial advantage. That is why it is one of the most favored architectural and monumental stone.

- Portland stone has been used in the construction of St. Paul's Cathedral, Buckingham Palace, Westminster palace, British Museum, the Bank of England.

Buckingham palace.

## Building Interior Decoration

Marble is recrystallized and then used for decoration of building interior. It is also used for statues, table surfaces and novelties.

Sandstone.

## Façade Construction

- Portland stone can be used for the construction of facades of reinforced concrete buildings.

- Conglomerate used in decoration.

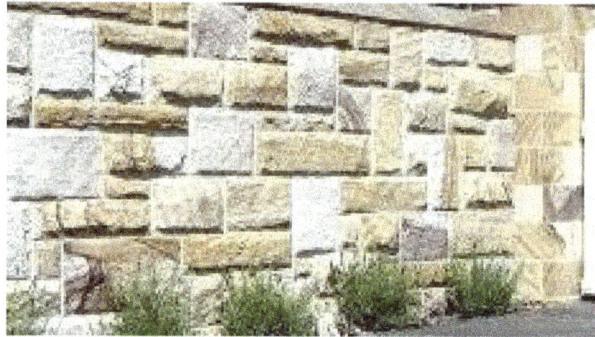

Exterior wall decoration.

- Filler in Paint: Shale which is a type of sedimentary rock that can be employed as a filler in many paints

- Sheetrock and Plaster: Gypsum is a major ingredient in sheetrock and plaster.

- Wallboard: Gypsum is ground up and employed for the production of wallboard.

## Others

Rail road ballast.

Sedimentary rocks are used in the construction of earthen dam, railroad ballast, canals, and as a rock fill.

# Permissions

# Index

**A**
Alluvial Fans, 2, 40, 98, 101-102, 104, 173-174, 187
Amber, 94, 104, 142
Arenites, 97-98, 107, 109

**B**
Beer Stone, 134
Biostratigraphy, 11, 23, 25, 29, 180
Biozones, 24
Blast Furnaces, 134
Breccia, 35, 50, 93, 96-99, 187

**C**
Calcium Carbonate, 5-6, 11-12, 14, 16, 19, 42, 99, 104, 126, 129, 132, 155, 157
Carbonate Platform, 31-38
Carbonate Precipitation, 31-32
Carbonate Sediment, 34-35, 37
Chemical Weathering, 9, 46-47, 49, 89, 110
Cherts, 45, 93, 123, 153, 155
Chronostratigraphy, 25-26
Clastic Rocks, 19, 97, 137
Clastic Sedimentary Rocks, 19, 42, 50, 93-94, 96-98, 105
Clay Matrix, 98, 109-110
Conglomerate, 29, 49-50, 62, 93, 96-100, 102, 177, 198
Coral Reefs, 5, 38, 132
Cross-bedding, 57, 59, 66-69, 76, 86, 99, 108, 148

**D**
Dolomites, 35-36, 39, 45, 93, 123, 127, 156

**E**
Euphotic Zone, 32, 37
Evaporite, 129, 135-140

**F**
Feldspar, 47, 97-98, 102, 104, 106, 108-110, 113-115, 121
Flue-gas Desulfurization, 134
Fluvial Environments, 2, 39, 59, 101
Fossil Content, 25, 29-30, 44

**G**
Glacial Margin, 41
Glauconite, 37, 125, 147

Graptolites, 23, 25
Guyots, 38, 125
Gypsum, 9, 16, 19, 45, 94, 103, 107, 135-136, 138-140, 192, 198

**H**
Halite, 9, 19, 94, 135-141
Highstand Shedding, 37-38

**I**
Igneous Rocks, 22, 25, 28, 47, 131
Index Fossils, 23-25
Indiana Limestone, 134
Iron Oxide, 47, 88, 94, 99, 104, 129
Isotopic Dating, 29-30

**J**
Jasper, 93, 103, 128, 155

**L**
Limestone, 15-16, 19-20, 31, 42, 47, 66, 93-94, 103, 114, 120, 123, 125-134, 139, 142, 147, 153, 157, 163, 193-197
Lithic Fragments, 106-107, 109-110
Lithification, 16, 35, 38, 65-66, 96, 99, 112, 156, 162
Lithostratigraphic Unit, 29, 31

**M**
Metamorphic Rocks, 22, 25, 47, 88-89, 109, 122
Minerals Calcite, 47, 126, 138
Mixed Layer, 12, 33
Mudrocks, 87, 97, 111-113, 116-120

**N**
Natural Gas, 21-22, 119-120, 122, 156, 192-193

**O**
Ooid, 16

**P**
Pelagic Clay, 42
Phosphate Minerals, 122, 124
Phosphorite, 7, 122, 124-126
Portland Cement, 127, 195
Principle Of Superposition, 18, 27, 62
Pyroxene, 106

## Q
Quartzite, 103

## R
Radioactive Decay, 22
Radiometric Dating, 19, 22, 30
Relative Sea Level, 37-38, 161, 175, 180
Rock Debris, 1, 44
Rock Salt, 7, 16, 94, 139, 141
Rock Strata, 22-25

## S
Sandstone Beds, 30
Sedimentary Facies, 20
Sedimentary Rocks, 2, 16-20, 22, 25, 28, 42, 46, 48, 50, 62-64, 86-89, 92-94, 99, 105-106, 111-112, 122-123, 127-128, 132, 135, 141, 149, 160, 163, 182, 189, 194, 196, 198
Seismic Activity, 103
Sequence Stratigraphy, 18, 37, 43, 159, 175, 178, 180-181, 184
Shale Rock, 120
Siliciclastic Sedimentary Rocks, 19, 111
Slope Shedding, 37-39
Stratification, 46, 64-69, 76, 84, 86-87, 89, 112, 115
Stromatolites, 33, 123

## T
Tidal Flat, 34, 87, 124, 158
Travertine, 16, 94, 127-129
Turbidity Currents, 4-5, 42, 55, 63, 81, 171

## U
Unconformity, 27, 101, 167
Uranium, 20, 83, 192

## V
Volcanic Activity, 6, 21, 123
Volcanic Rocks, 106, 109-110, 149

## W
Wave Base, 32-33, 35, 86-87, 183
Weathering, 1, 6-9, 17, 19, 46-49, 65-66, 69, 89, 92-94, 102, 106, 109-112, 114, 116, 118, 123, 154, 156, 191, 197

## X
Xenoliths, 28

## Z
Zircon, 106, 192

www.ingramcontent.com/pod-product-compliance
Lightning Source LLC
Chambersburg PA
CBHW082017190326

41458CB00010B/3216